AF564648

PROBLEMS IN PHYSICAL CHEMISTRY

PROBLEMS IN PHYSICAL CHEMISTRY

By
Dr. Shardendu Kislaya

DISCOVERY PUBLISHING HOUSE PVT. LTD.
NEW DELHI-110 002

First Published - 2011
Reprinted - 2017

ISBN: 978-81-8356-815-9

Problems in Physical Chemistry

Published by:
DISCOVERY PUBLISHING HOUSE PVT. LTD.
4383/4B, Ansari Road Darya Ganj
New Delhi - 110 002 (India)
Phone: +91-11-23279245, 43596064-65
Fax: +91-11-23253475
E-mail: discoverypublishinghouse@gmail.com
sales@discoverypublishinggroup.com
web: www.discoverypublishinggroup.com

Printed at:
Infinity Imaging Systems
Delhi

Preface

The book "**Problems in Physical Chemistry**" has been written to meet the requirements of graduate students of all Indian Universities. The subject matter of this book has been presented in this book has complete theory and large number of solved examples. We have been selected sufficient problems from various Universities examination papers.

I have spared no pains in applying my long experience of degree class teaching to make the book useful to the students. We have tried to our best to keep the book free from the misprint. The author shall be grateful to the readers who point out errors and omissions which inspite of all care might have been there.

The author in general hope. That the present book will be warmly received by the students and teachers. We shall indeed be very thankful to our colleague for their recommendations this book of for their students.

—Author

CONTENTS

1

ADVANCED PROBLEMS OF THE CHEMICAL EQUILIBRIUM

INTRODUCTION

When chemical equilibrium is established in one phase—a mixture of gases, a liquid solution - we have a case of *homogeneous equilibrium.* When more than one phases are involved—for example, gas and solid, or liquid and solid or liquid and gas equilibrium is said to be *heterogeneous.*

If a chemical equilibrium consists of two or more than two phases one of which at least must be a solid or liquid phase. It is said to be heterogeneous equilibrium. For example, water in contact with its vapour in a closed space forms a two phase (Liquid-Gas) heterogeneous equilibrium. A saturated solution in equilibrium with solute is another familiar instance of *heterogeneous equilibrium.*

Dissociation of red lead can similarly be regarded as a heterogeneous equilibrium consisting of three phases, two of which (Pb_3O_4 ad PbO) are solid and the third (O_2) a gas.

$$2Pb_3O_4(s) \rightleftarrows 6PbO(s) + O_2(g)$$

For the present we shall concern ourselves with gas-solid reactions which are the ore common.

(i) Thermal Decomposition of $CaCO_3$

A well known example of such a reaction is the dissociation of calcium carbonate:

$$CaCO_3(s) \rightleftarrows CaO(s) + CO_2(g)$$

where there is one gas and two solids. We could write the equilibrium constant as,

$$K_C = \frac{[CaO][CO_2]}{[CaCO_3]}$$

We now have to decide what is meant by the "concentrations" of the calcium carbonate and calcium oxide. These two solids will form two separate phases, since it is unusual for solids to mix. Any pure solid at a particular temperature and pressure, will contain a constant number of molecules per unit volume, so that the molar concentration of a solid is a constant.

Looking at it another way, we have seen that the vapour pressure of a liquid in a mixture is proportional to its molar concentration; the vapour pressure of pure liquid is constant, at a definite temperature, so that the concentration must be a constant. A solid also has a vapour pressure although it may be too small to be measured.

This vapour pressure is constant provided temperature is constant, irrespective of the amount of solid that may be present. By analog, we can use the vapour pressure of a solid as a measure of the concentration and we shall then arrive at the same result, *i.e.*, the pure solid has a constant concentration. The actual value that is given to these constants does not matter provided that we always use the same values. By convention, the *concentration of a pure liquid or a pure solid is taken as unity*. The actual amount present has no effect on the concentration.

The equilibrium constant for the calcium carbonate reaction now becomes.

$$K_C = [CO_2]$$

and since we are now dealing with the concentration of a gas, it is more convenient to use the pressure:

$$K_P = p_{CO_2}$$

The pressure of the CO_2 is also the total pressure of the system, since it is the only gas present. This is referred to as the *dissociation pressure* and since it is equal to K_P it follows that the dissociation pressure has definite value for any particular temperature.

The pressure of the system is fixed and cannot be altered without upsetting the equilibrium. Experimentally it has been found that the dissociation pressure of CO_2 in equilibrium with solid $CaCO_3$ and solid CaO, at any temperature, is constant.

Similar to homogeneous equilibria, the law of chemical equilibrium is also applicable to heterogeneous equilibria. The application of this law to the heterogeneous equilibria is based of the assumption that the reaction goes on i a heterogeneous system. Consider the above mentioned thermal dissociation of calcium carbonate.

$$CaCO_3(s) \rightleftarrows CaO(s) + CO_2(g)$$

We can assume that th reaction here is taking place between calcium carbonate vapour, calcium oxide vapour and carbon dioxide gas, the whole system is thus functioning like a homogenous system.

This assumption the leads us to further conclusion, that the solid $CaCO_3$ and the solid CaO, like other constituents also possess a definite vapour pressure and hence the equilibrium concentrations of the reacting substances can be expressed in terms of their partial pressures as in the case of other homogeneous equilibria. Therefore, p_{CaCO_3}, p_{CaO} and p_{CO_2}, respectively represent the partial pressures of $CaCO_3$, CaO and CO_2, then application of the law of chemical equilibrium gives the expression:

$$K_P = \frac{p_{CaO} \times p_{CO_2}}{p_{CaCO_3}}$$

The partial pressure of a solid (or a pure liquid) is very small but is constant at a particular temperature and remains so as ion as there is any solid (or liquid) present in the system p_{CaCO_3}, and p_{CaO} are thus constant. The equilibrium law equation, thus reduces to,

$$p_{CO_2} = K'_P$$

Thus, when calcium carbonate is heated in a closed vessel at a definite temperature, the pressure of CO_2 formed remains constant irrespective of the amount of calcium carbonate taken.

Therefore at a given temperature, the pressure of CO_2 will be constant. The equilibrium constant K'_P, therefore, appears to be determined only by th pressure of CO_2.

If CO_2 is allowed to escape, its pressure will decrease and further dissociation of calcium carbonate will take place until the original pressure is regarded. That is why in the line kilns, where the CO_2 escapes, the reaction proceeds practically in one direction, *i.e.*, the dissociation of $CaCO_3$. The equilibrium pressure of CO_2 which is constant at a particular temperature is known as the dissociation pressure of $CaCO_3$. The dissociation pressure at a particular temperature will to change, even if any product of the reaction is added or removed from the system. It changes with the change of temperature.

(ii) Thermal Dissociation of Solid Ammonium Hydrrosulphide

The solid ammonium hydrosulphide, on heating dissociates into two gases, viz., ammonia and hydrogen sulphide $NH_4HS(s) \rightleftarrows NH_3(g) + H_2S(g)$. It is another example of heterogeneous equilibrium. Since the partial pressure of solid component, *i.e.*, NH_4HS will remain constant at a particular temperature, as ion as there is any solid left in the system th equilibrium constant KP can be expressed as:

$$K_P = \frac{p_{NH_3} \times p_{H_2S}}{p_{NH_4HS}} = \frac{p_{NH_3} \times P_{H_2S}}{\text{constant}}$$

where p_{NH_3} and p_{H_2S} represent the partial pressures of ammonia and hydrogen sulphide. Evidently

$$K'_P = p_{NH_3} \times p_{H_2S}$$

i.e. At any temperature the product of the partial pressures of ammonia and hydrogen sulphide (p_{NH_4HS}) being very small, is neglected), and hence each will be equal to p/2.

i.e. $p_{NH_3} \times p_{H_2S} = p/2$

$$K'_P = p_{NH_3} \times p_{H_2S} = p/2 \times p/2 = p^2/4$$

Thus, from the total pressure of the system which is constant at a given temperature, the value K'_P can be determined. If one of the two gases is introduced in the system, from outside at equilibrium, p_{NH_3} will no longer be equal to p_{H_2S}, but the product $p_{NH_3} \times p_{H_2S}$ will still remain constant. Consequently, the dissociation of the solid will be diminished and a certain amount of solid hydrosulphide will be deposited.

Isambart confirmed the above observations by investigating the dissociation equilibrium of ammonium hydrosulphide and ammonium cyanide.

(iii) Iron-Steam Reaction

Another type of heterogeneous equilibrium is the reaction of steam with iron:

$$3Fe(s) + 4H_2O(g) \rightleftarrows Fe_3O_4(s) + 4H_2(g)$$

The equilibrium constant is given by

$$K_c = \frac{[Fe_3O_4][H_2]^4}{[Fe]^3 \times [H_2O]^4}$$

If the active asses of solids are taken constant, then

$$K' = \frac{[H_2]^4}{[H_2O]^4}$$

or
$$K'' = \frac{[H_2]}{[H_2O]}$$

or
$$K_P = \frac{p_{H_2}}{p_{H_2O}}$$

Experimentally also it has been found that the ratio p_{H_2}/p_{H_2O} is constant at any one temperature and independent of amounts or nature of iron and iron oxide present.

(iv) Water Gas Reaction

In this reaction steam is passed over heated coke:

$$H_2O(g) + C(s) \rightleftarrows CO(g) + H_2(g)$$

for which the equilibrium constant is

$$K_P = \frac{p_{CO} \times p_{H_2}}{p_{H_2O}}$$

as the partial pressure of carbon is taken as unity.

(v) Dehydration Equilibrium of Salt Hydrates

Dehydration of salts is an important class of heterogeneous reactions. In such reactions, the equilibrium constant is totally dependent of the vapour pressure of water. Salt hydrates during dehydration often dissociate in steps to form a number of intermediate hydrates according to the prevailing pressure of moisture is contact with th solid hydrates. Thus copper sulphate pentahydrate on dissociation yields trihydrate, monohydrate and then the anhydrous salt in the above order. Following three equilibria are established:

$$CuSO_4 . 5H_2O \rightleftarrows CuSO_4 . 3H_2O + 2H_2O \quad \text{(1st stage)}$$

$$CuSO_4 . 3H_2O \rightleftarrows CuSO_4 . H_2O + 2H_2O \quad \text{(2nd stage)}$$

$$CuSO_4 . H_2O \rightleftarrows CuSO_4 + H_2O \quad \text{(3rd stage)}$$

However, all the three equilibria do not exist simultaneously. Each one of them has an equilibrium pressure for its existence at a particular temperature. At 25°C the equilibrium vapour pressure for pentahydrate-trihydrate system is 7.8 mm, for the trihydrate-monohydrate, and monohydrate-anhydrous system, the pressure values are 5.6 mm and 0.8 mm respectively.

Application of law of chemical equilibrium to th equilibria for th three stages also proves that the equilibrium vapour pressures for three systems are independent of the active masses (or partial pressures) of the solids which are very small and are constant at a particular temperature and remain so, as long as there is any solid present in the system and are arbitrarily fixed as unity. They are, however, different, for different equilibria.

The equilibrium vapour pressure is not that of water vapour or any hydrate. It is the property of the system as a whole and hence, if the system changes, so must the equilibrium vapour pressure.

Thus, the equilibrium for the reaction at the *given temperature* is given by $K_{P_1} = p_{H_2O}^{2\ (1)}$ or $p_{H_2O}(1)m = \sqrt{K_{P_1}}$. So as long as $p_{H_2O}^{(1)}$ is equal to $\sqrt{K_{P_1}}$ both $CuSO_4 \cdot 5H_2O(s)$ and $CuSO_4 \cdot 3H_2O(s)$ will be present. When $P_{H_2O}^{(1)}$ becomes less than $\sqrt{K_{P_1}}$, more and more of $CuSO_4 \cdot 5H_2O(s)$ is converted into $CuSO_4 \cdot H_2O$ till the conversion is complete.

At this stage the third reaction starts for which the equilibrium constant is given by, $K_{P_3} = p_{H_2O}^{2\ (3)}$

Once again when $p_{H_2O}^{(3)}$ is less than K_{P_3} at a given temperature the dehydration reaction (in 3rd stage) takes place.

If isothermal dehydration of $CuSO_4 \cdot 5H_2O$ at 25°C is carried out, dehydrate is formed according to the 1st stage and the pressure remains constant so long as both penta and trihydrate are i equilibrium. Since the water vapour is being continuously removed, the pentahydrate will completely dissociate into trihydrate. Subsequent changes in this system on isothermal dehydration. Pressure 7.8 mm as shown in this figure corresponds to complete conversion of pentahydrate into trihydrate.

With the disappearance of pentahydrate, 1st stage equilibrium loses it significance and hence the pressure suddenly drops to 5.6 mm with the appearance of monohydrate. The pressure is constant so long as trihydrate and monohydrate are in equilibrium. Conversion of trihydrate into monohydrate is complete and hence the pressure again drops to 0.8 mm with the appearance of anhydrous copper sulphate.

The pressure would now fall only when monohydrate completely changes over to anhydrous copper sulphate. The stepwise process of dehydration of $CuSO_4 . 5H_2O$. During summer many salt hydrates *e.g.*, $FeSO_4 . 7H_2O$, $Na_2S_2O_3 . 5H_2O$ get easily dehydrated as K_p at high temperature is large and due to low humidity in air p_{H_2O} is less which are favourable for dehydration. All dehydration reactions can be reversed if the vapour pressure of water beyond a limit determined equilibrium pressure of water is increased.

(vi) Dissociation of Ammoniates of Silver Chloride or Deammination

This also provides another example of heterogeneous equilibria. The deammination of triammoniate of silver chloride AgCl, $3NH_3$ takes place through stages in the following successive equilibria:

$$\underset{\text{Triammoniate of silver chloride}}{2[AgCl . 3NH_3(s)]} \rightleftarrows \underset{\text{Sesqui-ammoniate of silver chloride}}{2AgCl . 3NH_3(s) + 3NH_3(g)}$$

$$2AgCl . 3NH_3(s) \rightleftarrows \underset{\text{Mono–ammoniate of silver chloride}}{2[AgCl . NH_3](s) + NH_3(g)}$$

$$AgCl . NH_3(s) \rightleftarrows AgCl(s) + NH_3(g)$$

This is similar to dehydration of hydrates and hence the values of the three equilibrium constants, may be represented as

$$Kp_1 = p_3NH_3 \quad \text{or} \quad pNH_3 = 3\sqrt{Kp_1}$$

$$Kp_2 = pNH_3 \quad \text{or} \quad pNH_3 = Kp_2$$

$$Kp_3 = pNH_3 \quad \text{or} \quad pNH_3 = Kp_3$$

During a given stage of demmination of ammoniates, the pressure of ammonia is constant and changes only when one particular stage of deammination is completed and the next stage sets in.

(vii) Liquid-Gas Equilibrium

Action of steam on molten sulphur represents an equilibrium of this type:

$$3S(1) + 2H_2O(g) \rightleftarrows SO_2(g) + 2H_2S(g)$$

The equilibrium in this case will be given by

$$\frac{(p_{SO_2}) \times (p_{H_2S})^2}{(p_{H_2O})^2} = K'p. \text{ [Sulphur, being pure liquid, } p_s = \text{constant].}$$

(viii) Solid-Liquid Equilibrium

Equilibrium between pure liquid and solid phases only is not possible, as the active masses in these cases have been assumed to be constant. Presence of a gaseous phase or a solution is necessary for any chemical equilibrium to be established.

DYNAMIC NATURE OF CHEMICAL EQUILIBRIUM

A chemical equilibrium can be defined as a *dynamic state at which the rate of forward and backward reactions are the same and at that instant all the reactants and products are present.* When chemical equilibrium is reached both the forward and backward reactions continue but at the same ate so that the concentrations of the components remain unchanged with time. Chemical equilibrium may be homogeneous or heterogeneous depending upon whether the reaction takes place :

(i) In one phase or

(ii) In two or ore phases.

Problems:

(i) When stream is passed over iron in a closed vessel, it will attain the state of equilibrium and the reactants and products will have a definite concentration at a given temperature.

$$2FE + 4H_2O \rightleftarrows Fe_3O_4 + 4H_2$$

(ii) If a mixture of hydrogen and iodine vapour is passed over platinum heated to 450°C in a closed vessel, the formation of hydrogen iodide and its decomposition into hydrogen and iodine takes place with equal and opposite rate and a state of equilibrium is reached.

$$H_2 + I_2 \rightleftarrows 2HI.$$

A reversible reaction must attain the equilibrium state. For instance, if hydrogen and iodine are taken in a reaction vessel, only toward reaction takes place as initially there is no hydrogen iodide. As soon as hydrogen iodide is formed, the reverse reaction starts. As time passes, the concentrations of hydrogen and iodine start decreasing and, therefore, the rate of forward reaction decreases.

However, the rate of reverse reaction increases. As time passes, an equilibrium state may be reached when the rates of toward and reverse reactions become équal. At this stage no further change in consecrations occurs.

Similar results would be obtained if we start with HI. At equilibrium, concentrations of HI, H_2 and I_2 become constant at a certain temperature. This condition is known as chemical equilibrium.

RECOGNITION OF ATTAINMENT OF EQUILIBRIUM

It is possible to recognize the attainment of equilibrium by constancy of certain observable properties. For example, the attainment of equilibrium is recognised in the evaporation of water at a given temperature by observing the pressure had become constant. Another example is the dissociation of calcium carbonate in which the attainment of equilibrium is recognised by observing that pressure of carbon dioxide had become constant. In both the examples, it becomes possible to recognise the attainment of equilibrium by observing constancy of pressure.

In some cases, the measurement of concentration of one or more of the reactants or products can be used to recognise the attainment of equilibrium. In order to understand this, we will consider the reaction between hydrogen gas and iodine vapour.

$$H_2(g) + I_2(g) \rightleftarrows 2HI(g)$$

Suppose one mole of hydrogen gas and one mole of iodine vapour are mixed in a flask of 1 litre capacity. Now suppose this flask is allowed to stand in a boiling bath (448°C) for several hours. If the reaction goes to completion, one should expect the formation of two moles of hydrogen iodide. But it is observed that the reaction comes to a stop when only 1.56 moles of hydrogen iodide get formed.

This reveals that only 0.78 mole of each of hydrogen and iodide have reacted. Thus, it is evident that 0.22 mole of hydrogen and 0.22 mole of iodine are still present as such without reacting. If the mixture is kept at the same temperature even for a couple of days, there will occur no change in the concentration of either two products or any of the reactants. In this reaction, the attainment of equilibrium is recognised by observing constancy of concentration.

A well known equilibrium reaction between $N_2O_4(g)$ and $NO_2(g)$ is considered,

$$\underset{\text{Almost colourless gas}}{N_2O_4(g)} \rightleftarrows \underset{\text{Deep reddish gas}}{2NO_2}$$

In this reaction, the colour of the gaseous mixture becomes constant at equilibrium.

From the above discussion, it can be concluded that the attainment of equilibrium can be recognised by observing constancy of properties such as pressure, concentration, density of colour whichever may be found to be suitable in a given case.

SYNTHESIS OF HYDROGEN IODIDE

The synthesis of hydrogen iodide can be represented as follows:

$$H_2 + I_2 \rightleftarrows 2HI.$$

Bodnstin selected this reaction for the experimental verification of law of mass action because it proceeds very slowly at ordinary temperatures but rapidly at the boiling point of sulphur (444°C) to establish the equilibrium. H_2 and I_2 mixtures of varying compositions are filled in glass bulbs and then sealed. These glass bulbs are heated in the vapour of boiling sulphur (Fig. 1) for sufficient time to attain equilibrium.

The sealed glass bulbs are then suddenly cooled to the room temperature to fix the equilibrium. The bulbs are then broken i the NaOH solution which absorbs the HI formed and I_2 and leaves H_2 free. Thus, the concentration of HI and I_2 are measured.

Suppose 'a' moles of H_2 and 'b' moles of iodine be heated in a sealed glass bulb having a volume V litres in a thermostat till equilibrium is established. If the concentration of HI formed after analysis be 2x, then according to the above reaction 2x moles of HI will be obtained from x moles of each H_2 and I_2.

The equilibrium concentration per litre of various reactants and product may be but as follows:

$$[HI] = 2x/V$$

$$[H_2] = (a - x)/V$$

$$[I_2] = (v - x)/V$$

Applying the law of mass action to the synthesis of HI

$$H_2 + I_2 \rightleftarrows 2HI$$

$$K = \frac{[HI]^2}{[H_2][I_2]}$$

Substituting the values of various terms

$$K = \frac{(2x/V)^2}{\left(\frac{a-X}{V}\right)\left(\frac{b-x}{V}\right)} = \frac{4x^2}{(a-x)(b-x)} \quad ...(1)$$

The two unknown factor sin the above equation are K and x and therefore, the law of mass action can be verified in two ways:

(a) The value of x is obtained by analysis and substituted in equation (1). If the value of K is always the same with different compositions of H_2 and I_2 mixtures the law of mass action is correct.

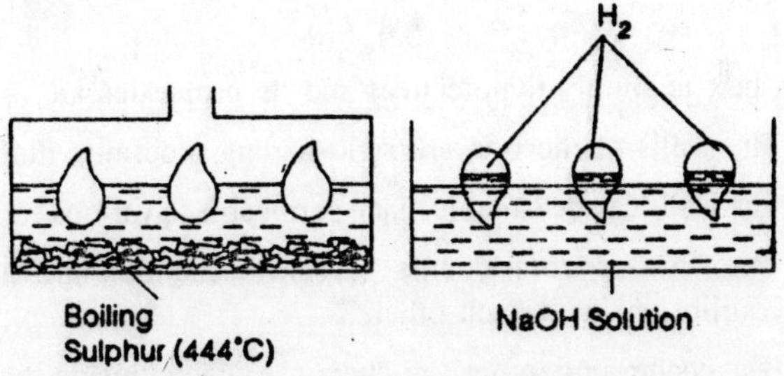

Fig. 1 : Experimental verification of law of mass action.

(b) From equation (1)

$$K = \frac{4x^2}{(a-x)(b-x)}$$

or $$ab - ax - bx + x^2 = \frac{4x^2}{K}$$

or $$x^2\left(1 - \frac{4}{K}\right) - x(a+b) + ab = 0$$

$\therefore$ $$x = (a+b) \pm \frac{\sqrt{(a+b)^2 - 4(1-4/K)(ab)}}{2(1-4K)}$$

From the values of K calculated from accurate observation, x can be calculated and if it agrees with the experimental values of x, th law is valid.

LAW OF MASS ACTION

Van't Hoff derived law of mass action thermodynamically by using an equilibrium box in which all the systems wherein equilibrium with each other at constant temperature (Fig. 2).

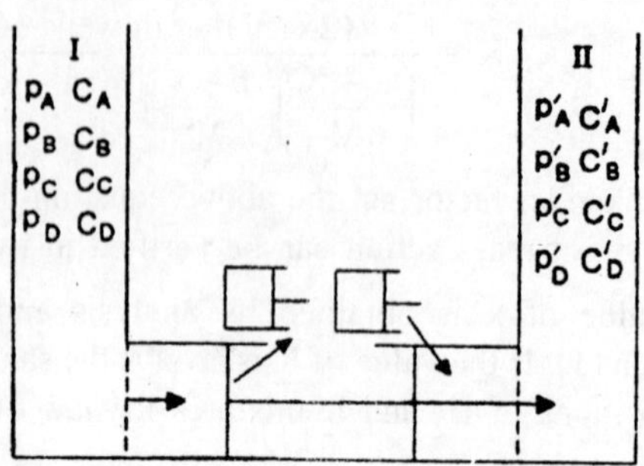

Fig. 2

This box is simply hypotectical and its properties are as follows:

(i) The walls of the box are rigid having indefinite dimensions.

(ii) The walls can be suitable made permeable to various components.

(iii) The various components present i the box are always in equilibrium with each other.

(iv) If a system undergoes the chemical change inside the box, the work done is zero.

(v) The box is assumed to have large dimensions so that there occurs no change in concentration or in partial pressure of any component, when a small amount of these is added or removed from it.

(vi) When the working substances are gases, they are considered to be ideal ones.

(vii) The box is kept at constant temperature.

Consider the following chemical change which is taking place at constant temperature.

$$A + B \rightleftarrows C + D$$

Consider two equilibrium boxes I and II in which components A, B, C, and D are always in equilibrium.

For box No (I) C_A, C_B, C_C and C_D represent the concentrations of A, B, C and D and p_A, p_B, p_C and p_D are corresponding partial pressures, similarly, C'_A, C'_B, C'_C and C'_D represent the concentrations of A, B, C and D and p_A, p_B, p_C and p_D the corresponding pressures in the box No. (II). Suppose one mole of the substance A gets transferred from (I) to box (II).

This can be achieved if it is assumed that th walls of the two boxes are preamble to A.

Now calculate the work done for the transfer of one mole of A which involves the following steps:

(a) One mole of A is removed from box No. I by moving the piston infinitesimally slowly outwards so that the partial pressure of A (*i.e.*, p_A) remains constant. The amount of work done is $P_V V_A$ where V_A represents the molar volume under pressure p_A.

(b) As the partial pressure of A is different in two boxes the gas removed from box I is changed isothermally and irreversibly from p_A and p'_A by working the piston, the amount of work done is

$$RT \ln \frac{p_A}{p'_A}$$

(c) The gas is now ready for putting it in box II. This will be put in box II by compressing the as infinitesimally slowly by work in the piston so that the partial pressure of A in box No. II (*i.e.* p'_A) remains constant. The work done in this process $p'_A V'_A$ where V'_A represents the molar volume of A under a partial pressure p'_A. Hence the total work done w_1 in transferring one mole of A reversibly and isothermally from box (I) to (II) is

$$w_1 = P_A V_A + RT \ln \frac{p_A}{p'_A} - p'_A V'_A$$

The gas is ideal, $p_A V_A = p'_A V'_A$

Hence $$w_1 = RT \ln \frac{p_A}{p'_A} \quad ...(1)$$

But the partial pressures are proportional to the molar concentrations. Therefore the amount of work done is

$$w_1 = RT \ln \frac{C_A}{C'_A} \quad ...(2)$$

Similarly, the work done w_2 in transferring one mole B of from box I to box (II) is as follows:

$$w_2 = RT \ln \frac{p_B}{p'_B} \quad ...(3)$$

$$w_2 = RT \ln \frac{C_B}{C'_B} \quad ...(4)$$

The concentrations of A and B do not undergo any change in box (I) as proportionate amounts of C and D will react to form A and B to maintain equilibrium.

Now one mole of C is removed from box (II) to box (I), then the amount of work done is

$$w_3 = RT \ln \frac{p'_C}{p_C} \quad ...(5)$$

$$w_3 = RT \ln \frac{C'_C}{C_C} \quad ...(6)$$

Also one mole of D is removed from box (II) to box (I). Then, the amount of work done w_4 is given by

$$w_4 = RT \ln \frac{p'_D}{p_D} \quad ...(7)$$

$$w_4 = RT \ln \frac{C'_D}{C_D} \quad ...(8)$$

When C and D are transferred to box (I), they interact to form necessary amounts of A and B to maintain the equilibrium.

This process can be considered as an isothermal cyclic one because

1. Both the boxes have regained their initial states and
2. The process is carried out at constant temperature.

For such an isothermal cyclic process, total amount of work done is zero, *i.e.*

$$w_1 + w_2 + w_3 + w_4 = 0$$

On substituting eqs. (1), (3), (5) and (7) in (9), we obtain

$$RT \ln \frac{p_A}{p'_A} + RT \ln \frac{p_B}{p'_B} + RT \ln \frac{p'_C}{p_C} + RT \ln \frac{p'_D}{p_D} = 0$$

or $$\ln \frac{p_A}{p'_A} + \ln \frac{p_B}{p'_B} = \ln \frac{p_C}{p'_C} + \ln \frac{p_D}{p'_D}$$

$$\ln \frac{p_A \times p_B}{p'_B \times p'_B} = \ln \frac{p_C \times p_D}{p'_C \times p'_D}$$

or $$\frac{p_A \times p_B}{p'_B \times p'_B} = \frac{p_C \times p_D}{p'_C \times p'_D}$$

or $$\frac{p'_C \times p'_D}{p'_A \times p'_B} = \frac{p_C \times p_D}{p_A \times p_B} = K_P \quad ...(10)$$

Similarly
$$\frac{C_C \times C_D}{C_A \times C_B} = \frac{C'_C \times C'_D}{C'_A \times C'_B} = K_C \quad ...(11)$$

[Using equations (2), (4), (6) and (1)]

Thus the equations (10) and (11) are the two forms of the law of mass action.

LAW OF MASS ACTION FROM CHEMICAL POTENTIAL

Consider a reaction

$$pA + qB + ... \rightleftarrows rC + mD + ...$$

Suppose the concentrations of the reactions A, B ... at the start of reaction are C_A, C_B ... respectively. Let the concentrations of the products C and C, ... at the end of reaction be C_C and C_D ... respectively. The free energies per mole (*i.e.*, chemical potentials) of the various substances at temperature T will be denoted by μ_A, μ_B, μ_C, μ_D ... respectively.

We know that the free energy of a system consist in of several components at constant temperature and pressure is give by

$$G = n_1\mu_1 + n_2\mu_2 + ... \quad ...(1)$$

where n_1, n_2 ... are the number of moles of the various components and m_1, m_2, ... are their respective chemical potentials. Hence one may write

Free energy of products = $r\mu_C + m\mu_D$

Free energy of reactants = $p\mu_A + q\mu_B$

$$\therefore \quad \Delta G = (r\mu_C + m\mu_D) - (p\mu_A + q\mu_B) \quad ...(2)$$

The chemical potential of a substance in any state is give by the equation

$$\mu = \mu^o + RT \ln a$$

where μ^o is chemical potential in the standard state of unit activity and μ the chemical potential at activity a.

Replacing the values of chemical potentials of the products and the reactants in equation (2), we get

$$\Delta G = [r(\mu_C^o + RT \ln a_C) + m(m_D^o + RT \ln a_D)] + ...$$

$$-[p(\mu_A^o + RT \ln a_A) + q(\mu_B^o + RT \ln a_B) + \ldots]$$

Rearranging,

$$\Delta G = [(r\mu_C^o + m\mu_D^o + \ldots) - (p\mu_A^o + q\mu_B^o + \ldots)] + RT \ln \frac{(a_C)^r (a_D)^m \ldots}{(a_A)^p (a_B)^q \ldots} \quad \ldots(4)$$

The μ^o terms refer to standard states of unit activity of the substance concerned. The first term in square brackets on the right side of equation (4), therefore, represents the increase in free energy of the reactants when these at unit activity react to form products also at the unit activity. If this energy increase is represented by ΔG^o, equation (4) may be written as

$$\Delta G = \Delta G^o + RT \ln \frac{(a_C)^r (a_D)^m \ldots}{(a_A)^p (a_B)^q \ldots} \quad \ldots(5)$$

$$= \Delta G^o + RT \ln x \quad \ldots(6)$$

where x stands for reaction quotients of activities of the products and reactants, viz.,

$$x = \frac{(a_C)^r (a_D)^m \ldots}{(a_A)^p (a_B)^q \ldots}$$

Equation (6) is commonly known as the reaction isotherm.

If the above reaction is in equilibrium state, then ΔG will be zero. Hence from equation (5),

$$\Delta G^o = -RT \ln \left(\frac{(a_C)^r (a_D)^m}{(a_A)^p (a_B)^q} \right) \quad \ldots(7)$$

Since ΔG^o represents standard free energy change, *i.e.*, when the reactants as well as products are in their standard states of unit activity it must be constant at a given temperature. Consequently the expression on the right side of equation (7) must be constant as well. Since R is a gas constant, it follows, therefore, that if the temperature T is constant, then

$$\left(\frac{(a_C)^r (a_D)^m}{(a_A)^p (a_B)^q} \right) = \text{Constant} = K_a \quad \ldots(8)$$

where K_a is the equilibrium constant of the reaction. Equation (8) is the mathematical statement of the law of mass action.

Substituting equation (8) in (7), we get

$$\Delta G^o = -RT \text{ in } K_a \qquad ...(9)$$

This is another important relation.

REACTION ISOTHERM OF THE VAN'T HOFF'S

This can be deduced by either of the following methods:

1. Free Energy Change Method

In the earlier deduction it was assumed that the system remains in equilibrium and hence the change in free energy is zero. Now let us proceed to calculate the change in free energy when a chemical reaction is taking place out at constant temperature from some arbitrary concentrations of reactants to some other arbitrary concentrations of the products. Let us assume a reaction involving for gases, A, B, C and D

$$A + B \rightleftarrows C + D \qquad ...(1)$$

I this reaction amounts of A and B are decreasing while those of C and D are increasing. Subsequently, the free energies of A and B are decreasing while those C and D are increasing.

The free energy of the substance A per mole at temperature T is give as follows:

$$G_A = G_A^o + RT \ln p_A$$

where p_A represents the pressure of A and G_A^o represents the free energy at some standard state (p = 1) and is known as standard free energy.

Similarly, the free energies of B, C and D are as follows:

$$G_B = G_A^o + RT \ln p_B$$

$$G_C = G_C^o + RT \ln p_C$$

$$G_D = G_D^o + RT \ln p_D$$

By definition, the free energies of the reaction (ΔG) is

$$\Delta G = (\Delta G_C + G_D) + (G_A + G_B)$$

Hence

$$\Delta G = (G_C^o + RT \ln p_C + G_D^o + RT \ln p_D) - (G_A^o + RT \ln p_A + G_B^o + RT \ln p_B)$$

$$= (G_C^o + G_D^o - G_A^o - G_B^o) + RT \ln \frac{p_C \times p_D}{p_A \times p_B}$$

$$= \Delta G^o + RT \ln \frac{p_C \times p_D}{p_A \times p_B} \quad ...(2)$$

where ΔG^o represents standard free energy of reaction.

But at equilibrium $\Delta G = 0$

Therefore, $\Delta G^o + RT \ln \left[\frac{p_C \times p_D}{p_A \times p_B}\right] = 0$

where pressure are equilibrium pressures,

or $\Delta G^o + RT \ln K_p = 0$

or $\Delta G^o = -RT \ln K_p$

where $K_p = \dfrac{p_C \times p_D}{p_A \times p_B}$

On substituting this value of ΔG^o in equation (2), we get

$$\Delta G = -RT \ln K_P + RT \ln \frac{p_C \times p_D}{p_A \times p_B}$$

or

$$-\Delta G = RT \ln K_P - RT \ln \frac{p_C \times p_D}{p_A \times p_B} \quad ...(3)$$

Now consider the general reaction

$$n_1A + n_2B + ... \rightleftarrows^* n_3C + n_4D + ...$$

By following the same procedure, one may write

$$-\Delta G = RT \ln K_P + RT \ln \frac{(p_C)^{n_3} \times (p_D)^{n_4}}{(p_A)^{n_1} \times (p_B)^{n_2}} \quad ...(4)$$

$$-\Delta G = RT \ln K_P - RT \Sigma \ln p \quad ...(5)$$

Thus $-\Delta G = RT \ln K_c - RT \Sigma \ln c \quad [\because \; p \propto c] \quad ...(6)$

where K_c is equilibrium constant. Equations (5) and (6) are known as Van't Hoff isotherm, because the reaction is taking place at constant temperature. This relation is very important because at may be used to show the direction in which a reaction tends to proceed.

2. The Van't Hoff's Equilibrium Box Method

It is a theoretical device which has been used successfully for calculating the free energy change in a chemical reaction taking place at constant temperature from arbitrarily chosen starting concentration of the reactants up to some of other arbitrarily chosen concentrations of the reactants. Consider the following chemical reaction.

$$n_1A + n_2B \rightarrow n_3C + n_4D \quad ...(1)$$

Let there be four reservoirs (Fig. 3) each of indefinite dimensions, one for each of the components A, B, C and D. Let the concentration, pressure and volume of A in the reservoir I be C_A, P_A and V_A, respectively. Similarly let C_B, P_B and V_B, C_C, P_C and VC and C_D, P_D and V_D be the corresponding concentrations, pressures and volumes of components B, C and D i the reservoirs II, I and IV respectively. Besides these vessels, there is a central reservoir which is an equilibrium box V having A, B C and D where their concentrations, partial pressures and molar volumes are $(C_A)_e$, $(P_A)_e$ and $(V_A)_e$, $(C_B)_e$, $(P_B)_e$ and $(V_B)_e$ $(C_C)_e$, $(P_C)_e$ $(P_C)_e$ and $(C_D)_e$, $(P_D)_e$ and $(V_D)_e$ respectively. Let us assume the temperature to be T. Now transfer n_1 moles of A from reservoir I into the equilibrium box V. This can be reversibly done by the following three stage process:

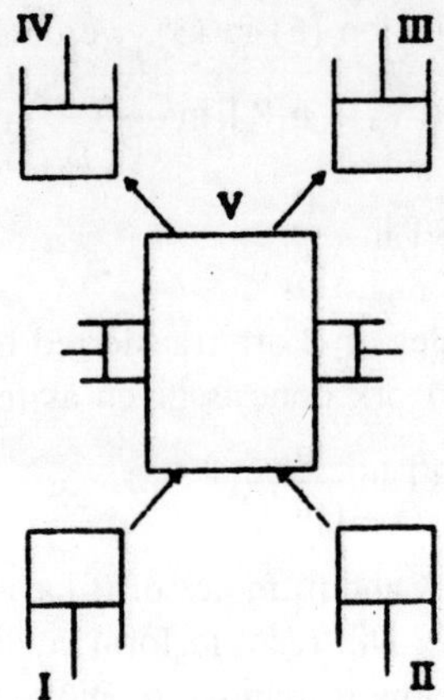

Fig. 3

(1) The n_1 moles of A at pressure P_A and temperature T are taken out by slowly moving a piston of reservoir I. Thus, the necessary amount of A will come out through a wall which is permeable to A. It will involve work done by the gas, equivalent to n_1, P_A, V_A, *i.e.*

$$(\text{work done})_1 = n_1(P_AV_A) \quad ...(2)$$

(2) Before the gas has been introduced into the equilibrium reservoir

V, the pressure of A is changed from P_A to $(P_A)_e$. The work done in the process for n_1 moles is

$$(\text{work done})_2 = n_1 RT \ln \frac{P_A}{(P_A)_e} \qquad ...(3)$$

(3) In the last stage, the gas is compressed at a pressure of $(P_A)_e$ into the equilibrium box. This involves work done which is equal to

$$(\text{work done})_3 = -n_1 (P_A)_e (V_A)_e \qquad ...(4)$$

The total work done is, therefore, equal to sum of the equations (2), (3) and (4), *i.e.*,

$$w_1 = (\text{work done})_1 + (\text{work done})_2 + (\text{work done})_3$$

or $$w_1 = n_1 P_A V_A + n_1 RT \ln \frac{P_A}{(P_A)_e} - n_1 (P_A)_e (V_A)_e \qquad ...(5)$$

But the gas is ideal and the process is isothermal. Therefore,

$$n_1 P_A v_A = n_1 (P_A)_e (V_A)_e \qquad ...(6)$$

On substituting equation (6) in (5), we obtain

$$w_1 = n_1 P_A V_A + n_1 RT \ln \frac{P_A}{(P_A)_e} - n_1 P_A V_A$$

$$= n_1 RT \ln \frac{P_A}{(p_A)_e} \qquad ...(7)$$

Similarly, if n_2 moles of B are transferred from reservoir II to the equilibrium box V the work done is given as follows:

$$w_2 = n_2 RT \ln \frac{p_B}{(p_B)_e} \qquad ...(8)$$

When n_1 moles of A and n_2 moles of B have bee introduced inside the equilibrium box they will react to form equilibrium concentrations of C and D. We have now to remove n_3 moles of C and n_4 moles of D from the equilibrium box V into the respective reservoirs III and IV. The work w_3 ad w_4 involved in these processes are as follows:

$$w_3 = n_3 RT \ln \frac{(p_C)_e}{p_C} \qquad ...(9)$$

$$w_4 = n_4 RT \ln \frac{(p_D)_e}{p_D} \qquad ...(10)$$

Total work done in this process is given as follows:

$$W = w_1 + w_2 + w_3 + w_4$$

$$W = n_1 RT \ln \frac{P_A}{(P_A)_e} + n_2 RT \ln \frac{P_B}{(P_B)_e}$$

$$+ n_3 RT \ln \frac{(P_C)_e}{P_C} + n_4 RT \ln \frac{(P_D)_e}{P_D} \quad ...(11)$$

The work W is derived from the free energy decrease of the system *i.e.*

$$W = -\Delta G \quad ...(12)$$

From equations (11) and (12), we have

$$-\Delta G = n_1 RT \ln \frac{P_A}{(P_A)_e} + n_2 RT \ln \frac{P_B}{(P_B)_e}$$

$$+ n_3 RT \ln \frac{(P_C)_e}{P_C} + n_4 RT \ln \frac{(P_D)_e}{P_D} \quad ...(13)$$

$$-\Delta G = RT \ln \frac{(P_C)_e^{n_3} (P_D)_e^{n_4}}{(P_A)_e^{n_1} (P_B)_e^{n_2}} - RT \ln \frac{(P_C)^{n_3} (P_D)^{n_4}}{(P_A)^{n_1} (P_B)^{n_2}}$$

$$= RT \ln K_P - RT \Sigma n \ln P$$

Where $K_P = \frac{(P_C)_e^{n_3} (P_D)_e^{n_4}}{(P_A)_e^{n_1} (P_B)_e^{n_2}}$

and $\Sigma \ln n P = \ln(P_C)\, n_3 + \ln(P_D) n_4 - (\ln P_A)\, n_1 - (\ln P_B) n_2$

If, however, the work done is expressed in terms of C instead of P we would get

$$-\Delta G = RT \ln \frac{(C_C)_e^{n_3} (C_D)_e^{n_4}}{(C_A)_e^{n_1} (C_B)_e^{n_2}} - RT \ln \frac{(C_C)^{n_3} (C_D)^{n_4}}{(C_A)^{n_1} (C_B)^{n_2}}$$

$$= RT \ln K_c - RT \Sigma n \ln C \quad ...(15)$$

where the term $\Sigma n \ln C$ has a similar significance as $\Sigma n \ln P$ already referred to.

Van't Hoff's Reaction Isochore

Van't Hoff's reaction isochore deals with the variation of equilibrium constant with temperature. According to Van't Hoff's reaction isotherm, the change in free energy is given as follows:

$$\Delta G = RT \ln K_c - RT \Sigma \ln C \quad ...(1)$$

The Van't Hoff's isochore can be derived by differentiating equation (1) of Van't Hoff's reaction isotherm with respect to temperature at constant volume. But at constant volume the change in free energy is not given by ΔG but by ΔA. Therefore, equation (1) modifies to

$$-\Delta A = RT \ln K_c - RT \Sigma \ln c \qquad ...(2)$$

On differentiating it with respect to temperature at constant volume, we obtain

$$\left(\frac{\partial(-\Delta A)}{\partial T}\right)_V = RT\frac{d}{dT}(\ln K_c) + R \ln K_c$$

$$-RT\frac{d}{dT}(\Sigma n \ln c) + R \ln c$$

$RT\frac{d}{dT}\Sigma_n \ln c = 0$ because this arbitrary concentration is not considered as a function of temperature. Thus,

$$\left(\frac{\partial(-\Delta A)}{\partial T}\right)_V = RT\frac{d}{dT}(\ln K_c) + R \ln K_c - R \Sigma n \ln c$$

or $$-\frac{\partial(\Delta A)}{\partial T} = RT\frac{d}{dT}(\ln K_c) + R \ln K_c - R \Sigma n \ln c$$

or $$-\frac{\partial(\Delta A)}{\partial T} = RT\frac{d}{dT}(\ln K_c) + R \ln K_c - R \Sigma n \ln c$$

or $$-T\frac{\partial(\Delta A)}{\partial T} = RT^2\frac{d}{dT}(\ln K_c) + R \ln K_c - R \Sigma n \ln c$$

$$= RT^2\frac{d}{dT}(\ln K_c) - \Delta A \qquad ...(3)$$

[use equation (2)]

The Gibb's Helmholtz equation is as follows:

$$\Delta A = \Delta E + T\left[\frac{\partial(AA)}{\partial T}\right]_V$$

or $$-T\left[\frac{\partial(AA)}{\partial T}\right]_V = \Delta E - \Delta A \qquad ...(4)$$

On comparison of equation (3) with (2), we get

$$\Delta E = RT^2\frac{d}{dT}\ln K_c$$

or $$\frac{\Delta E}{RT^2} = \frac{d}{dT}\ln K_c \qquad ...(5)$$

But $q_v = \Delta E$ where q_v is the heat absorbed by the system at constant volume. Hence we get

$$\frac{d}{dT}\ln K_c = \frac{q_v}{RT^2} \qquad ...(6)$$

It is the required equation.

Integrated Form of the Van't Hoff Isochore

The integrated form of this can be obtained by integrating equation (5) between the temperatures T_1 and T_2 then

$$\int_1^2 d\ln K_c = \int_{T_1}^{T_2} \frac{\Delta E}{RT^2}\, dT$$

$$\frac{K_{c_2}}{K_{c_1}} = \frac{\Delta E}{R}\left[\frac{1}{T_1} - \frac{1}{T_2}\right] \qquad ...(7)$$

This equation is useful for calculating the value of equilibrium constant at any temperature provided equilibrium constant at one temperature and ΔE are given. Even ΔE can also be calculated when K_{c_1} and K_{c_2} at two temperature T_1 and T_2 are known.

REACTION ISOBAR OF THE VAN'T HOFF

Some confusion exists in the literature about the reaction isobar. Some prefer to call it by the name of Van't Hoff reaction Isotherm. We will now deduce this equation by using the following relation.

$$K_P = K_c(RT)^{\Delta n}$$

where Δn represents change in number of moles during the reaction.

On taking logarithms of both sides, we get

$$\ln K_P = \ln K_c + \Delta n \ln RT$$

or

$$\ln K_P = \ln K_c + \Delta n \ln R + \Delta n \ln T$$

On differentiating this equation with respect to temperature we obtain

$$\frac{d}{dT}(\ln K_P) = \frac{d}{dT}(\ln K_c) + \frac{d}{dT}(\Delta n \ln R) + \frac{d}{dT}(\Delta n \ln T)$$

But $\frac{d}{dT}[\Delta n \ln R] = 0$, R is a constant

$$\therefore \quad \frac{d}{dT}(\ln K_P) = \frac{d}{dT}(\ln K_c) + \frac{\Delta n}{T} \qquad \left(\because \frac{d}{dT}\ln T = \frac{1}{T}\right)$$

But $\frac{d}{dT}(\ln K_c) = \frac{q_v}{RT^2}$ [From reaction isochore]

$$\therefore \quad \frac{d}{dT}(\ln K_c) = \frac{q_v}{RT^2} + \frac{\Delta n}{T} = \frac{q_v + \Delta n\, RT}{RT^2} \quad ...(1)$$

For an ideal as,

$$P\Delta V = \Delta n\, RT$$

Therefore, equation (1) becomes as follows:

$$\frac{d}{dT}(\ln K_P) = \frac{q_v + P\Delta V}{RT^2}$$

But $q_p = q_v + P\Delta V$, where q_p represents the heat absorbed by the system at constant pressure. Hence,

$$\frac{d}{dT}(\ln K_P) = \frac{q_p}{RT^2} \quad ...(2)$$

We also know

$$\Delta H = q_p$$

$$\therefore \quad \frac{d}{dT}(\ln K_P) = \frac{\Delta H}{RT^2} \quad ...(3)$$

This equation is known as Van't Hoff's reaction isobar. By integrating equation (3), we obtain

$$\int_1^2 d \ln K_P = \int_{T_1}^{T_2} \frac{\Delta H}{RT^2}\, dH$$

$$\frac{K_{P_2}}{K_{P_1}} = \frac{\Delta H}{R}\left[\frac{1}{T_1} - \frac{1}{T_2}\right]$$

Knowing the equilibrium constant K_{P_2} at temperature T_2 it is possible to calculate the equilibrium constant K_{P_1} at temperature T_1 and *vice versa*, provided the heat of reaction ΔH is known. Alternatively if the equilibrium constants of a reaction at two temperatures are known, the heat of reaction (ΔH) can be calculated.

APPLICATIONS OF MASS ACTION

1. Dissociation of Hydrogen iodide

This is first type (*i.e.*, $K_P = K_c$) of a homogeneous gaseous reversible reaction as in this case the total number of molecules on both sides is the same.

	$2HI$ ⇄*	H_2 +	I_2 – Heat
Initial	1 mole	0	0
At equilibrium	$(1-x)$ mole	$x/2$ mole	$x/2$ mole
Concentration	$\frac{(1-x)}{V}$	$\frac{x}{2V}$	$\frac{x}{2V}$

(where V = volume of vessel)

Let us start with '1' g mole of HI enclosed in a vessel of volume V litres. Let 'x' mole of HI decompose at the equilibrium to yield x/2 mole of each H_2 and I_2. Then the equilibrium molar concentrations are:

$$[HI] = \frac{(1-x)}{V};$$

$$[H_2] = \frac{x/2}{V}; \quad [I_2] = \frac{x/2}{V}$$

∴ From the Law of Mass action, we get

$$K_P = K_C = \frac{k_f}{k_b} = \frac{[H_2][I_2]}{[HI]} = \frac{\frac{x}{2V} \cdot \frac{x}{2V}}{\left[\frac{1-x}{V}\right]^2} = \frac{x^2}{4(1-x^2)}$$

(1) *Effect of pressure change,* From the above equation it is clear that K_c is *independent* of total volume 'V' of the system. A change of pressure alters the total volume and as K_c is independent of 'V' consequently *change of pressure will not alter the final state of equilibrium.*

(2) *Effect of temperature,* With the rise and fall of temperature the velocity constants k_f and k_b vary differently. It has been found that k_f increases more rapidly than k_b with the increase of temperature. So, with the increase of temperature $K_c = k_f/k_b$ will increase. Evidently, the concentration of products (*i.e.* H_2 and I_2) will increase and that of reactant HI will decrease. *Hence, the temperature will favour the dissociation of HI.*

(3) *Effect of adding products,* Alternative equation for equilibrium constant K_P in terms of partial pressures, will be:

$$K_P = \frac{P_{H_2} \times P_{I_2}}{P_{HI}^2}$$

On adding of any product, say H_2 the value of P_{H_2} will increase. In order to keep the value of K_P constant, the value of P_{I_2} must decrease and of P_{HI} must increase. In other words, *adding of H_2 will suppress the dissociation of HI.* Similar will be the consequences if I_2 is added.

(4) *Effect of adding an inert gas:* Addition of an inert gas like argon to the equilibrium mixture will increase the total pressure of the reactants without change the partial pressures of H_2, I_2 or HI. Hence, *the addition of an inert gas will have no effect on the degree of dissociation of HI.*

2. Dissociation of PCl_5

It is a second type (*i.e.*, $K_P \neq K_c$) of a homogeneous gaseous reaction as the total number of molecules of the reactants *change* as a result of chemical reaction.

$$PCl_5 \rightleftharpoons^* PCl_3 + Cl_2$$

	PCl_5	PCl_3	Cl_2
Initial	1 mole	0	0
At equilibrium	(1 – x) mole	x mole	x mole
		= 1 – x + x + x = 1 + x mole	
Concentration	$\frac{(1-x)}{V}$	$\frac{x}{V}$	$\frac{x}{V}$
			where V = Volume
Partial pressure	$\frac{(1-x)}{(1+x)}P$	$\frac{xP}{(1+x)}$	$\frac{xP}{(1+x)}$
			where P = Pressure

Let us start with '1' g mole of PCl_5 enclosed in a vessel of volume 'V' litres. Let 'x' mole of PCl_5 decompose at the equilibrium to yield x mole of each PCl_3 and Cl_2. Then, equilibrium molar concentrations are:

$$[PCl_5] = \frac{1-x}{V}$$

$$[PCl_3] = [Cl_2] = \frac{x}{V}$$

$\therefore$ Applying Law of Mass Action we get

$$\{K_c = \frac{k_f}{k_b} = \frac{[PCl_3][Cl_2]}{[PCl_5] = \frac{x^2}{(1-x)V}} \quad ...(i)$$

The expression for K_P can be deduced as follows:

Total No. of moles at equilibrium = 1 – x + x + x = (1 + x). Let P = Total pressure of the system,

Then $\quad P_{PCl_5} = \frac{1-x}{1+x}P; \quad P_{PCl_3} = \frac{xP}{1+x};$

$$P_{PCl_2} = \frac{xP}{1+x}$$

$$\therefore \quad K_P = \frac{p_{PCl_3} \times p_{Cl_2}}{p_{PCl_5}}$$

$$= \frac{x^2P}{(1-x)(1+x)} = \frac{x^2P}{(1-x^2)} \quad ...(ii)$$

(1) *Effect of pressure change:* In the expression (ii) for K_P, p is in the numerator. Hence, it follows if P is increased, the value of 'x' must decrease i order to keep K_P constant. Evidently increase in pressure suppresses the dissociation of PCl_5.

(2) *Effect of temperature:* It has been found that *k_f increases more rapidly than k_b with the rise of temperature.* Thus, it is evident that with the increase of temperature the value of $K = k_f/k_b$ increases, *i.e.*, the concentration of products increases and that of reactant PCl_5 decreases. *Hence, high temperature favours the dissociation of PCl_5.*

(3) *Effect of adding products at constant volume:* We have:

$$K_P = \frac{p_{PCl_3} \times p_{Cl_2}}{p_{PCl_5}}$$

Addition of any product, say Cl_2 will result in the increase in the value of PCl_3. In order to keep K_P constants, the value of PCl_3 will decrease and that of PCl_5 would decrease. Evidently, *addition of Cl_2 suppresses the dissociation of PCl_5.* Similarly consequences will happen by addin PCl_3.

(4) *Effect of adding an inert gas at constant volume:* The addition of an inert gas will increase the total pressure P, since K_P is proportional to P. Hence, in order to keep K_P constant the increase of P by the addition of inert gas will decrease the value of x. Evidently, *addition of an inert as will support the dissociation of PCl_5.*

3. Dissociation of N_2O_4

This is a second type (*i.e.*, $K_P \neq K_C$) of homogeneous gaseous reaction as the number of molecules alters as a result of reaction.

	$N_2O_4 \rightleftarrows^*$	$2NO_2$
Initial	1 mole	0
At equilibrium	$(1-x)$ mole	2x mole

Total = 1 – x + 2 = 1 + x moles

Concentration $\frac{(1-x)}{V}$ $\frac{2x}{V}$ where V = Volume

Partial pressure $\frac{(1-x)}{(1+x)}P$ $\frac{2xP}{(1+x)}$ where P = Pressure

Let us start with 1 mole of N_2O_4 enclosed in a vessel of capacity 'V' litres. Let at equilibrium 'x' mole of N_2O_4 is dissociated into 2x moles of NO_2. Then, equilibrium concentrations are:

$$[N_2O_4] = \frac{1-x}{V};$$

$$[NO_2] = \frac{2x}{V}$$

∴ Applying Law of Mass Action, we get

$$K_C = \frac{k_f}{k_b} = \frac{[NO_2]^2}{[N_2O_4]} = \frac{4x^2}{(1-x)V} \quad ...(i)$$

The expression for K_P can be calculated as follows:

Total No. of moles at equilibrium = 1 – x + 2x = 1 + x. Let total pressure of the system = P.

Then, $$p_{N_2O_4} = \frac{(1-x)}{(1+x)}P;$$

$$p_{NO_2} = \frac{2xp}{1+x}$$

$$\therefore \quad K_P = \frac{(p_{NO_2})^2}{p_{N_2O_4}} = \frac{4x^2P}{(1-x^2)} \quad ...(ii)$$

(1) *Effect of pressure change:* From the expression (ii) it is clear that when P increases x must decrease and *virsa* in order to keep K_P constant. In other words, *increase of pressure suppresses the dissociation of N_2O_4.*

(2) *Effect of adding product at constant volume:* We have

$$K_P = p^2_{NO_2} / p_{N_2O_4}$$

If NO_2 is added to the equilibrium mixture in a close vessel, the partial pressure p_{NO} will increase. In order to keep K_P constant, the value of $p_{N_2O_4}$ must increase. Hence, *dissociation of N_2O_4 will be suppressed by the addition of NO_2.*

(3) *Effect of change of temperature:* It has been found that with the

rise of temperature k_f increases more rapidly than k_b, i.e., K increases or the concentration of product NO_2 will increase. Hence, *increase in temperature will favour the formation of NO_2, i.e., dissociation of N_2O_4.*

(4) *Effect of adding an inert gas at constant volume:* The addition of an inert gas will increase the total pressure P, since K_P, is directly proportional to P. Hence, in order to keep K_P constant, the increase of total pressure P by the addition of inert gas will decrease the value of 'x'. Evidently, *the addition of an inert gas at constant volume will suppress the dissociation of N_2O_4.*

4. Synthesis of Ammonia

This is a second type (*i.e.*, $K_P \neq K_C$) of homogeneous gaseous reaction as the total number of molecules *changes* as a result of reaction

	N_2 +	$3H_2 \rightleftarrows^*$	$2NH_3$
Initial	1 mole	3 moles	0
At equilibrium	(1 – x) mole	(3 – 3x) moles	2x mole

= 1 – x + 3 – 3x + 2x = 4 – 2x

Concentration	$\frac{(1-x)}{V}$	$\frac{(3-3x)}{V}$	$\frac{2x}{V}$

where V = Volume

Partial pressure	$\frac{(1-x)P}{(4-2x)}$	$\frac{(3-3x)P}{(4-2x)}$	$\frac{2xP}{(4-2x)}$

where P = Pressure

Let us start with 1 mole of N_2 and 3 moles of H_2 in a closed vessel of volume 'V' litres. Let 'x' mole of N_2 and 3x moles of H_2 are used up at equilibrium. Then, equilibrium concentrations are:

$$[N_2] = \frac{1-x}{V}; \quad [H_2] = \frac{3(1-x)}{V}; \quad [NH_3] = \frac{2x}{V}$$

∴ Applying law of Mass Action, we get

$$K_C = \frac{k_f}{k_b} = \frac{[NH_3]^2}{[N_2][H_2]^3} = \frac{\left[\frac{2x}{V}\right]^2}{\left[\frac{1-x}{V}\right]\left[\frac{3(1-x)}{V}\right]^3} = \frac{4x^2V}{27(1-x)^4} \quad ...(i)$$

The expression for K_P can be calculated as follows:

Total No. of molecules in the system at equilibrium

$$= 1 - x + 3 - 3x + 2x = 4 - 2x.$$

Let total pressure of system = P.

Then

$$p_{N_2} = \frac{1-x}{4-2x}P;$$

$$p_{H_2} = \frac{3-3x}{4-2x}P;$$

$$p_{NH_3} = \frac{2xP}{4-2x}$$

$$\therefore \qquad K_P = \frac{(p_{NH_3})^2}{p_{N_2} \times (p_{H_2})^3}$$

$$= \frac{\left[\dfrac{2xP}{4-2x}\right]^2}{\left[\dfrac{1-x}{4-2x}P\right]\left[\dfrac{(3-3x)P}{4-2x}\right]^3} = \frac{16x^2(2-x)^2}{27P^2(1-x)^4} \qquad \text{...(ii)}$$

1. *Effect of pressure change:* As K_P is inversely proportional to P^2, so an increase in P will cause a decrease in x. In other words, *increase of pressure increases the formation of NH_3.*

2. *Effect of temperature change:* It has bee found that *with the rise of temperature k_b increases more rapidly than k_f.* Thus, with the increase in temperature the backward reaction is favoured. Hence, at high temperature the dissociation of NH_3 will occur. Conversely *low temperature favours the formation of NH_3.*

3. *Effect of adding reactants at constant volume:* We have:

$$K_P = \frac{(p_{NH_3})^2}{p_{N_2} \times (p_{H_2})^3} m$$

Addition of any reactant, say N_2, will result ion the increase in the value of p_{N_2}. In order to keep K_P constant, the value of p_{NH_3} will increase and that of p_{H_2} will decrease. Evidently, addition of *N_2 favours the formation of NH_3.* Similar consequences will happen by adding H_2.

4. *Effect of adding an inert gas at constant volume:* The addition of an inert gas will decrease the total pressure P, since K_P is inversely proportional to P.

Hence, in order to keep K_P constant the increase of P by addition of inert gas will increase the value of x. Evidently, *addition of an inert gas will favour the formation of NH_3.*

5. Formation of Nitric Oxide

This is a first type (*i.e.*, $K_P = K_C$) of a homogeneous reaction as the total number of molecules remains the same as a result of the reaction:

	N_2 +	O_2 ⇄*	2NHO – Heat
Initial	a mole	b moles	0
At equilibrium	(a – x) mole	(b – x) moles	2x mole
			= 1 – x + 3 – 3x + 2x = 4 – 2x
Concentration	$\frac{(a-x)}{V}$	$\frac{b-x}{V}$	$\frac{2x}{V}$

(where V = Volume of vessel)

Let us start with 'a' moles of N_2 and 'b' moles of O_2 in a closed vessel of capacity 'V' litres.

Let at equilibrium 'x' moles each of N_2 and O_2 are used up to form O_2 moles of NO. Then, the equilibrium concentrations are:

$[N_2] = (a - x)/V$;

$[O_2] = (b - x)/V$;

$[NO] = 2x/V$

∴ Applying Law of Mass Action, we get

$$K_P = K_C = \frac{[NO]^2}{[N_2][O_2]} = \frac{4x^2}{(a-x)(b-x)}$$

1. *Effect of pressure change:* As V ∝ 1/P and K_C is independent of V. Therefore, K_C is independent of P. Hence, *the equilibrium concentrations of reaction remain uneffected by change of pressure.*

2. *Effect of adding reactants:* We have

$$K_P = \frac{p^2_{NO}}{p_{N_2} \times p_{O_2}}$$

of adding a reactant, say N_2 the partial pressure of N_2 will increase. In order to keep K_P constant, the value of p_{O_2} will decrease and that of p_{NO} will increase, *i.e.*, more N_2 and O_2 will react to form NO.

In other words, addition of N_2 will favour the formation of NO. Similar effects will be caused by the addition of oxygen.

3. *Effect of temperature change:* It has been found that with the rise of *temperature* k_f *increases more rapidly than* k_b *i.e.* with the increase of temperature the about of product, *i.e.*, NO, will increase.

Hence *formation of nitric oxide is favoured by high temperature.*

6. Esterification of Ethyl Alcohol by Acetic Acid

This is a reaction in solution and is represented by the equation.

	CH_3COOH +	C_2H_5OH ⇄	$CH_3COOC_2H_5$ +	H_2O
Initial	a mole	b moles	0	0
At equilibrium	(a – x) mole	(b – x) moles	x mole	x mole
Concentration	$\frac{(a-x)}{V}$	$\frac{b-x}{V}$	$\frac{x}{V}$	$\frac{x}{V}$

Let us start with 'a' moles of acid and 'b' moles of alcohol. Let 'x' moles each of ester and water are formed at equilibrium. If V is thee total volume of the solution mixture, then equilibrium molar concentrations are:

$$[CH_3COOH] = (a - x)/V;$$

$$[C_2H_5OH] = \frac{b-x}{V};$$

$$[CH_3COOC_2H_3] = x/V \quad [H_2O] = x/V$$

∴ Applying Law of Mass Action, we get

$$K_C = \frac{k_f}{k_b}$$

$$= \frac{[CH_3COOC_2H_5][H_2O]}{[CH_3COOH][C_2H_5OH]}$$

$$= \frac{x/V \times x/V}{\frac{a-x}{V} \times \frac{b-x}{V}} = \frac{x^2}{(a-x)(b-x)}$$

CHARACTERISTICS OF CHEMICAL EQUILIBRIUM

Let us now consider certain important characteristics of chemical equilibrium. These are:

1. Chemical equilibrium, at a given temperature, can be characterised by constancy of certain observable properties like pressure,

concentration, density or colour.

2. Chemical equilibrium can be approached from either direction, let us take an example to illustrate this important characteristic of chemical equilibrium. We know that hydrogen gas and iodine vapour react at elevated temperature yielding hydrogen iodide gas, according to the equation.

$$H_2(g) + I_2(g) \rightarrow 2HI(g)$$

We know that hydrogen iodide is not a very stable compound. This on heating dissociates into hydrogen and iodine.

$$2HI(g) \rightarrow H_2(g) + I_2(g)$$

It means that the hydrogen-iodine reaction is reversible and can be represented as

$$H_2(g) + I_2(g) \rightleftarrows 2HI(g)$$

When 1 mole of H_2 is allow to mix with 1 mole of iodine in a closed vessel of one litre capacity and the vessel is allowed to stand for a long time at 448°C (the temperature of boiling sulphur), ultimately, only 1.56 moles of HI, gets formed. There occurs of further increase in the amount of HI even if the reaction is permitted to continue for several days at the same temperature of 448°C. This reveals that only 0.78 mole of each of H_2 and I_2 has reacted. The composition of the reaction mixture at equilibrium therefore, would be 1.56 moles of Hi, 0.22 mole of H_2 and 0.22 mole of I_2.

Now the reaction is carried out by taking 2 moles of Hi i the same vessel and keeping it again in the both of boiling sulphur at 448°C. After about the same time as before, it will be found that only 0.44 mole of HI gets dissociated into H_2 and I_2. The composition of the reaction mixture, therefore, would be 1.56 moles of HI, 0.22 mole of H_2 and 0.22 mole of I_2 as before.

This reveals that when equilibrium gets attained at a given temperature each reactant and each product is having a fixed concentration and this is independent of the fact whether the reaction is started with the reactants or with the products. Thus, equilibrium can be attained from either side.

3. A catalyst can hasten the approach of equilibrium but fails to alter the state of equilibrium. In other words, it means that the

relative concentrations of the products and the reactants remain the same irrespective of the presence or absence of a catalyst. For example,

$$2SO_2(g) + O_2(g) \rightleftarrows 2SO_3(g)$$

This reaction is allowed to take place, generally, at 500°C. The equilibrium in this case would be only attained after a long time. However, if a catalyst (platium or vanadium pentoxide) is add to the system, the equilibrium would be attained fairly rapidly but the equilibrium state is not altered. In other words, the relative amounts of SO_2 (the product) and SO_2 and O_2 (the reactants) at equilibrium would remain the same irrespective of the fact whether a catalyst is added or not in the reaction. Hence the yield of SO_2 would not be improved in any way by using the catalyst.

4. Chemical equilibrium is dynamic in nature because it involves two opposing reactions. One of these reactions takes place from the reactants towards the products and is known as the forward reaction. The other takes place from the products towards the reactants ad is known as the reverse reaction. When equilibrium is attained, there occurs no further change in the concentrations of the products or the reactants. This gives the idea that the reaction has stopped. But this is not the case. Actually, the opposing reactions, the forward reaction ad the reverse reaction, are taking place simultaneously at equal rates. Hence one reaction completely undoes the effect of the other. In other words, as much of the products get formed in a given time as change back into the reactants i the same time. Hence the chemical equilibrium is dynamic and not static

Le-chatelier's Principle : Prediciting Equilibrium State

This generalisation which predicts the influence on a dynamic equilibrium by altering concentration of species, total pressure and temperature was announced independently by Le-Chatelier in 1885 and Braun in 1886.

This is entirely a general law and applies to physical as well as chemical systems. It may be stated as, *When a system in equilibrium is subjected to a stress, i.e., change of concentration, pressure or temperature, the equilibrium shifts i such a direction as to undo or tends to undo the effect of the change or stress.*

This law is of great help in prediciting the conditions for getting maximum yields of the desired products which will be clear from the following considerations:

The Effect of Changing Concentrations of the Reactants

If any of the components at the equilibrium state is added from outside, its concentration increases ad hence, the equilibrium tends to change in such a way as to counteract the effect of increase of concentration of the added substance. In other words, the equilibrium will shift in the direction which the added substance is being consumed. This is illustrated by the following example:

Let us consider the following reaction in aqueous solution:

$$Fe^{3+}(aq) + SCN^{-}(aq) \rightleftarrows Fe(SCN)^{2+}(aq)$$

If soluble thiocyanate salt is added to the above equilibrium solution containing both Fe^{3+} (aq) and SCN^{-} the red colour of the complex ion $Fe(SCN)^{2+}$ increases. A new equilibrium is then attained in which more $Fe(SCN)^{2+}$ is present than that was there before the addition of SCN^{-} Increasing the concentration of SCN^{-} ions increases the concentration of the $Fe(SCN)^{2+}$. This is in accord with *Le-Chatelier's principle*. The change imposed on the system was an increase in the concentration of SCN^{-}. The change can be counteracted in part by some Fe^{3+} and SCN^{-} ions reacting to form $Fe(SCN)^{2+}$.

The same argument applies to addition of ferric ion from a soluble ferric salt. In each case, the formation of $Fe(SCN)^{2+}$ uses up a portion of the added reactant practically counteracting the change.

So in a chemical equilibrium, increasing the concentrations of the reactants shifts the equilibrium in the direction of the produces whereas increasing the concentrations of the products shifts the equilibrium in the direction of the reactants.

The Effect of Altering Pressure

In most of the system composed of liquids and solids, the effect of pressure on equilibrium is relatively shifts. In gaseous reactions, on the on the hand, the effect of pressure is often extremely important.

Increase of external pressure will cause the equilibrium to shift in the direction which will bring about a lowering of pressure. In a chemical reaction, this means that the equilibrium will shift in the direction which produces the smaller umber of as molecules. This is clear from the following equilibrium:

Example (i) $PCl_5(g) \rightleftarrows PCl_3(g) + Cl_2(g)$

On increasing the total pressure, the system will now occupy a much smaller volume than it did previously and the total umber of molecules present per unit volume is greater than it was under the original equilibrium conditions. This change can be counteracted in part if some PCl_3 and Cl_2 combine to form PCl_5. The total number of molecules present is reduced (2 molecule unite to form 1 molecule).

Therefore, the increase in pressure shifts the above equilibrium i the reverse direction.

(ii) $N_2(g) + 3H_2(g) \rightleftarrows 2NH_3(g)$

The increase in pressure causes the equilibrium to shift form left to right (forward reaction is favoured).

(iii) $\underset{\text{1 mole}}{N_2O_4(g)} \rightleftarrows \underset{\text{2 moles}}{2NO_2(g)}$

In this case also if the pressure is increased the volume of the system decreases ad the total number of moles per unit volume is now more than before. The change can be encountered if the equilibrium shifts in the direction in which total number of molecules decreases. Thus according to *Le-Chatelier's principle*, on applying pressure the system shifts i favour of N_2O_4, *i.e.*, suppresses the dissociation of N_2O_4.

(iv) $\underset{\text{2 moles}}{2SO_2(g)} + \underset{\text{1 mole}}{O_2(g)} \rightleftarrows \underset{\text{2 moles}}{2SO_3(g)}$

The increase in pressure tends to favour equilibrium in the forward direction, *i.e.*, the formation of sulphur trioxide.

(v) $\underset{\text{1 mole}}{CaCO_3(s)} \rightleftarrows \underset{\text{1 mole}}{CaCO(s)} + \underset{\text{1 mole}}{CO_2(g)}$

In this case increase in pressure favours the backward reaction, *i.e.*, suppresses the dissociation of $CaCO_3$.

(vi) $N_2(g) + O_2(g) \rightleftarrows 2NO(g)$

If the pressure on this equilibrium is increased, the gases are compressed to a smaller volume and the concentrations are all increased. Let us suppose that the equilibrium state is changed to favour products, *i.e.*, some N_2 and O_2 molecules react to form a exactly equal number of molecules of NO_2. Since no change takes place in the total number of moles, the proposed change in the equilibrium does not partially reduce the pressure change. According to Le-Chatelier's principle, the processes occur so as to counteract partially imposed change. So no change of the equilibrium state in this reaction is expected when the pressure is altered.

The equilibrium state is not affected by a pressure change for any equilibrium gas mixture where the number of reactant molecules is the same as the number of product molecules in the balanced reaction.

(3) The Effect of Altering Temperature

The shifting of an equilibrium by change of temperature can also be predicted by applying Le-Chatelier's principle. *If the temperature of a system i equilibrium is increased, the equilibrium will shift in the direction which absorbs heat.* Lowering of the temperature will shift the equilibrium in the direction i which heat evolves.

Example: (i) *In the reversible reaction,*

$$N_2(g) + 3H_2(g) \rightleftarrows 2NH_3; \Delta H = -92.38 \text{ kJ}$$

increase in temperature shifts the equilibrium from right to left (absorption of heat), and decrease in temperature will shift it form left to right (evolution of heat).

The equilibrium mixture will contain less ammonia at high temperature, and more at low temperature or formation of ammonia is not favoured at high temperature.

(ii) *In the reaction,*

$$N_2(g) + O_2(g) \rightleftarrows 2NO(g); \Delta H = +180.5 \text{ kJ}$$

increase in temperature will shift the equilibrium from left to right (absorption of heat), and decrease in temperature will shift it from right to left (evolution of heat) or formation of NO is favoured at high temperature.

(iii) *In the reaction,*

$$PCl_5(s) \rightleftarrows PCl_3 + Cl_2(g); \quad \Delta H = -153.72 \text{ kJ}$$

increase in temperature shifts the equilibrium from right to left (absorption of heat), and decrease in temperature will shift it from left to right (evolution of heat). Higher temperature thus favours dissociation of PCl_5.

(iv) *In the reaction,*

$$2SO_2(g) + O_2(g) \rightleftarrows 2SO_3(g); \quad \Delta H = -96.25 \text{ kJ}$$

increase i temperature shifts the equilibrium from right to left (absorption of heat) or favours the backward reaction. A decrease i temperature shifts the equilibrium from left to right (evolution of heat) or favours formation of SO_3.

(v) $CaCO_3(s) \rightleftarrows CaO(s) + CO_{2(g)}$;

$\Delta H = +179.96$ kJ

The decomposition of $CaCO_3$ is accompanied by absorption of heat (endothermic reaction), thus according to *Le-Chatelier's principle*, increase in temperature favours decomposition of $CaCO_3$.

Le-Chatelier's principle finds extensive application in predicting the coditions for getting maximum yield of products in reaction of industrial importance. A few common examples are discussed below.

(i) Formation of Ammoia

(Haber's Process)

$$\underset{\text{1 mole}}{N_2} + \underset{\text{3 moles}}{3H_2} \rightleftarrows \underset{\text{2 moles}}{2NH_3}; \quad \Delta H = -92.38 \text{ kJ}$$

(a) *Effect of temperature:* If temperature is raised, the yield of NH_3 diminishes and the proportion of N_2 and H_2 in the equilibrium mixture increases. On the other hand, when the temperature is lowered the equilibrium shifts i the direction i which heat is produced, *i.e.*, toward reaction. Hence, *low temperature favours the formation of ammonia.* It may be pointed out that at low temperature, the equilibrium is reached very slowly. Even with catalyst, the rate of reaction is not rapid at low temperature. Hence, a compromise is to be made between the low temperature, and time required for equilibrium to be reached. A temperature of 500°C is generally used.

(b) *Effect of pressure:* If at a constant temperature, pressure is increased the total volume will decrease but number of molecules per unit volume will increase. To counteracts that, equilibrium will shift in the direction of decrease of molecules *i.e.*, the forward direction. Hence *high pressure favours the formation of ammonia.*

(c) *Effect of increasing concentrations of reactants:* By adding a quality of one or both of the reactants (N_2 or H_2) to the equilibrium mixture, the reaction will proceed in the direction in which the add components are used up, *i.e.*, the forward direction. Hence, *high concentration of N_2 or H_2 (or both) favours the formation of NH_3.*

(ii) *Oxidation of SO_2 to SO_3*

(*Contact process for the manufacture of sulphuric acid*).

$$\underset{2\text{ moles}}{2SO_2} + \underset{1\text{ mole}}{O_2} \rightleftarrows \underset{2\text{ moles}}{2SO_3}\,;\ \Delta H = -96.25\text{ kJ}$$

The forward reaction is *exothermic* and is accompanied by a *decrease in number of moles.*

(a) *Effect of temperature :* If temperature is raised, then according to the Le-Chatelier's principle the heat supplied to the mixture constitutes an imposed stress ad reaction will take place to counter this stress, *i.e.*, the resulting reaction is such that it tends to lower the temperature. But a reaction that lowers temperature is endothermic, it is the reverse of the above exothermic reaction,

$2SO_3 \rightarrow 2SO_2 + O_2$ – Heat energy

In other words, as the temperature rises, the yield of SO_2 diminishes and the concentration not the reactants ($SO_2 + O_2$) increases.

(b) *Effect of concentration :* The addition of more of oxygen imposes a stress. According to Le-Chatelier's principle, to counteract the stress more SO_3 will be formed; thereby decreasing the concentration of SO_2.

(c) *Effect of pressure :* If a constant temperature, the pressures increases, the total volume will decreases but the total umber of molecules per unit volume will increase. To counteract that, equilibrium will shift in the direction of decrease of molecules, *i.e.*, the forward direction.

(iii) *For Nitric of Nitric Oxide*

(*Birkeland and Eyde Process*)

$$\underset{1\text{ mole}}{N_2(g)} + \underset{1\text{ mole}}{O_2(g)} \rightleftarrows \underset{2\text{ mole}}{2NO}\,;\ \Delta H = 180.5\text{ kJ}$$

The forward reaction is *endothermic and no change in volume occurs.*

(a) *Effect of temperature* : As the forward reaction is accompanied by a absorption of heat, cosequetly *a increase in temperature will favour the formation of NO.*

(b) *Effect of pressure* : As the toward reaction is accompanied by no change in volume, consequently the pressure will have no effect on the equilibrium.

(c) *Effect of concentration* : The formation of NO will be favoured b an excess of O_2 or N_2 or both.

Applications of Le-Chatelier's Principle to Physical Equilibria

(1) *Effect of pressure on melting point of ice* : At 0°C ice is i equilibrium with water and this equilibrium can be expressed in the form of an equation which is give below. Ice occupies more volume than corresponding amount of water at the same temperature.

$$\underset{\text{more volume}}{\text{ice}} \rightleftarrows \underset{\text{less volume}}{\text{water}}$$

If this system is contained in a cylinder with a piston of which the temperature is 0°C and if pressure be increased then according to Le-Chatelier's principle the system will adjust in the direction of the lesser volume, *i.e.*, more ice will melt. This means the melting point of ice is lowered by increasing the pressure and *vice versa.*

(2) *Effect of temperature on melting of ice*

$$\underset{\text{more volume}}{\text{ice}} \rightleftarrows \underset{\text{less volume}}{\text{water}} \quad - \text{Q joules or } \Delta H = +\text{ve}$$

The forward reaction is accompanied by absorption of heat. Hence increase of temperature favours the formation of water.

(3) *Effect of pressure on solubility of salts* : We come across two type of salts. Those which *dissolve with a decrease in volume* [*i.e.*, the volume of solution is less than the volume of the solid and volume of the solvent] their solubility according to this principle will increase with increase of pressure and *vice versa.*

And the second type is of those salts which dissolve leading to an increase in volume, their solubility will decease with increase of pressure and *vice versa.*

(4) *Effect of temperature on solubility of salts :* The dissolution of salts like KCl, NH_4Cl etc.,

$$NH_4Cl + aq \rightleftarrows NH_4Cl(aq) - Q \text{ joules or } \Delta H = +\text{ve}$$

is accompanied by absorption of heat and hence the forward reaction or solubility of these salts will increase with the increase of temperature.

On the other had, solubility of substances like NaOH, KOH, $Ca(OH)_2$ etc.

$$NaOH + aq \rightleftarrows NaOH\ (aq) + Q \text{ joules or } \Delta H = -\text{ve}$$

which are accompanied by evolution of heat, their solubility will decrease with the rise of temperature.

(5) *Effect of heat and pressure on vapourisation of water*

$$\text{Water (I)} \rightleftarrows \text{Water vapour} - Q \text{ joules or } \Delta H = +\text{ve}$$

The forward reaction is accompanied by a large increase of volume and absorption of heat. Hence an increase in pressure will favour the backward reaction, *i.e.*, liquefaction, while increase of temperature will favour the forward reaction, *i.e.*, vapourisation.

TREATMENT OF LE-CHATELIER'S PRINCIPLE

Let us consider the following general chemical reaction:

$$aA + bB \rightleftarrows mM + nN$$

As the free energy change, ΔG for the reaction depends upon the temperature, pressure and the extent of the reaction ζ, it implies that

$$\Delta G = f(T, P, \zeta) \qquad ...(1)$$

As ΔG is a state function, the total change in the reaction free energy with T, P and ζ can be put as follows:

$$\text{or } d(\Delta G) = \left(\frac{\partial(\Delta G)}{\partial T}\right)_{P,\zeta} dT + \left(\frac{\partial(\Delta G)}{\partial P}\right)_{T,\zeta} dP + \left(\frac{\partial(\Delta G)}{\partial \zeta}\right)_{T,P} d\zeta \qquad ...(2)$$

$$\text{But, } \Delta G = \left(\frac{\partial G}{\partial \zeta}\right)_{T,P} = G' \text{ (By definition)} \qquad ...(3)$$

Therefore, eq. (2) can be put as follows:

$$\therefore \quad d(\Delta G) = \frac{\partial}{\partial T}\left\{\left(\frac{\partial G}{\partial \zeta}\right)_{T,P}\right\}_{P,\zeta} dT$$

$$+\frac{\partial}{\partial T}\left\{\left(\frac{\partial G}{\partial \zeta}\right)_{T,P}\right\}_{T,\zeta} dP + \frac{\partial}{\partial \zeta}\left(\frac{\partial G}{\partial \zeta}\right) d\zeta \qquad ...(4)$$

From the mathematical properties of state functions, it is possible to express d(ΔG) as follows:

$$\therefore\ d(\Delta G) = \frac{\partial}{\partial \zeta}\left\{\left(\frac{\partial G}{\partial T}\right)_{P,\zeta}\right\} dT + \frac{\partial}{\partial \zeta}\left\{\left(\frac{\partial G}{\partial P}\right)_{T,\zeta}\right\} dT + \left(\frac{\partial^2 G}{\partial \zeta^2}\right) d\zeta \qquad ...(5)$$

But $(\partial G/\partial T)_P = -S$ and $(\partial G/\partial P)_T = V$. Therefore, Eq. (2) may be written as follows:

$$\therefore\ d(\Delta G) = -\left(\frac{\partial S}{\partial \zeta}\right)_{T,P} dT + \left(\frac{\partial V}{\partial \zeta}\right)_{T,P} dP + \left(\frac{\partial^2 G}{\partial \zeta^2}\right)_{T,P} d\zeta \qquad ...(6)$$

The quantity $(\partial S/\partial \zeta)_{T,P} = \Delta S$ = Rate of change of entropy with the progress variable, ζ *i.e.*, this quantity refers to the entropy change of the reaction and the quantity $(\partial V/\partial \zeta)_{T,P} = \Delta V$ refers to the volume change of the reaction. Thus, Eq. (5) can be put as follows:

$$d(\Delta G) = -\Delta S\, dT + \Delta V\, dP + \left(\frac{\partial^2 G}{\partial \zeta^2}\right)_{T,P} d\zeta \qquad ...(7)$$

If a case is considered where T, P and get changed in such a manner that equilibrium is maintained, then, ΔG would be minimum and d(ΔG) = 0, so that

or $$-\Delta S\ dT + \Delta V\ dP + G'' d\zeta = 0 \qquad ...(8)$$

where $G'' = (\partial^2 G/\partial \zeta^2)_{T,P}$ is positive because G is minimum at equilibrium.

At equilibrium, ΔG = 0 and ΔS = ΔH/T, so that

$$-(\Delta H/T)dT + \Delta V dP + G'' d\zeta = 0 \qquad ...(9)$$

Let us now consider two cases.

Case A: Variation of ζ with T at constant P at equilibrium. If P is constant, dP = 0, so that Eq. (9) becomes as follows:

$$-(\Delta H/T)(dT)_P + G'' d\zeta = 0 \qquad ...(10)$$

$$(\partial \zeta/\partial T)_P = \Delta H/(TG'')$$

As G″ is a positive quantity, the sign of $(\partial \zeta/\partial T)_P$ would depend on the sign of ΔH. Remember that ($\Delta H = H_{products} - H_{reactants}$).

(i) For isothermic reactions, ΔH is negative ($H_{products} < H_{reactants}$). Hence, $(\partial\zeta/\partial T)_P$ would be negative.

If dT is positive (which is the case when temperature gets increased), t~en $d\zeta$ would be negative, *i.e.*, ζ should decrease with increase in T and the concentration of the reactant would increase in the direction of high enthalpy.

When dT is negative (*i.e.*, when temperature gets decreased), $d\zeta$ should be positive, *i.e.*, ζ should increase with decrease in T and concentration of the product would get increased in the direction of low enthalpy.

(ii) For endothermic reactions, ΔH is positive ($H_{products} < H_{reactants}$). Hence $(\partial\zeta/\partial T)_P$ would be positive.

When dT is negative, (*i.e.*, when the temperature is lowered), then $d\zeta$ too, must be negative) so than ζ should decrease with decrease in T and reactants with lower enthalpy are formed.

If dT is positive (*i.e.*, when the temperature is increased), then $d\zeta$ should also be positive and ζ should increase causing the formation of more products with higher enthalpy.

Hence, it can be concluded that an increase in temperature would shift the equilibrium towards the high enthalpy side and a decrease in temperature would shift the equilibrium towards the low enthalpy side. This statement has been similar to a statement of Le Chatelier's principle.

Case B. Variation of ζ with P at constant T at equilibrium : When T is constant, dT = 0, so that Eq. (9) would become as follows:

$$\Delta V(dP)_T + G''(d\zeta) = 0 \quad ...(11)$$

$$(\partial\zeta/\partial P)_T = \Delta V/G'' \quad ...(12)$$

As G'' is positive, the sign of $(\partial\zeta/\partial T)_P$ would depend upon that of ΔV.

(i) When $\Delta V = 0$ (*i.e.*, $V_{products} = V_{reactants}$) then ζ would be independent of P at constant T.

(ii) When ΔV is positive (*i.e.*, $V_{products} > V_{reactants}$) the RHS of Eq. (12) would be negative so that $(\partial\zeta/\partial T)_P$ will also be negative.

Increase in pressure ($dP > 0$) will decrease $\zeta(d\zeta > 0)$ so that equilibrium would shift towards the reactant side. On the other hand, decrease in pressure ($dP < 0$) will increase $\zeta(d\zeta > 0)$ therefore equilibrium would shift towards the product side.

(iii) When ΔV is negative (*i.e.* $V_{products} < V_{reactants}$) then the RHS of Eq. 12 would be positive so that $(\partial\zeta/\partial P)_T$ would also be positive.

Increase in pressure ($dP > 0$) will increase $\zeta(d\zeta > 0)$ so that equilibrium will get shifted towards the product side. On the other hand decrease in pressure ($dP < 0$) will decrease $\zeta(d\zeta < 0)$, so that equilibrium would be shifting towards the reactant side.

Thus it can be concluded that an increase in pressure would shift the equilibrium towards the low volume side of the reaction, whereas a decease in pressure would shift the equilibrium towards the high volume side. This is another statement of Le Chatelier's Principle.

DE DONDERS CONCEPT OF DEGREE OF ADVANCEMENT OF A REACTION

In order to understand this concept, *i.e.*, the progress of a reaction, we shall consider the hypothetical reaction of the following type which is proceeding in a closed vessel.

$$aA + bB \rightleftarrows cC + dD.$$

As this reaction is being carried out in a closed vessel, it means that the total mass of the system, evidently, will remain constant although the number of moles of each reactant and product will go on changing with time. If ζ represents the progress or degree of advancement of the reaction then $d\zeta$ will represent the small increase in the extent of the reaction that takes place i any given interval of time. Due to this, the amount of A decreases by a $d\zeta$ and that of B by b $d\zeta$ while the amount of C increases by cd ζ and that of D increased by d $d\zeta$. Mathematically, these changes may be represented as follows:

$$dn_A = -ad\zeta \qquad dn_B = -bd\zeta$$

$$dn_C = -cd\zeta \qquad dn_D = -dd\zeta$$

If the temperature and pressure of the system are assumed to be constant, the change in Gibbs free energy for this small progress of this reaction will be as follows:

$$(dG)_{T,P} = \sum_{l} \mu_i dn_i$$

$$= (c\mu_C + d\mu_D - a\mu_A - b\mu_B)d\zeta$$

or $$\left(\frac{\partial G}{\partial z}\right)_{T,P} = (c\mu_C + d\mu_D - a\mu_A - b\mu_B)$$

$$= \sum_i v_i \mu_i$$

where v_i is the difference between the stoichiometric coefficients of the products and those of the reactants, *i.e.*, (c + d) – (a + b).

De Donder introduced the term $\left(\frac{\partial G}{\partial \zeta}\right)_{T,P}$ which may be defined as the change of *Gibbs free energy per unit change in the extent of the reaction.* Three cases may arise:

(i) If the value of this term is negative, the reaction will proceed from left to right.

(ii) If the value of this term is positive, the reaction will proceed from right to left.

(iii) If the value of this term is zero, th system will be in a state of equilibrium.

DERIVATION OF LAW OF MASS ACTION

From Kinetic Theory of Gases

According to kinetic considerations the number of reactions per second between any tow gases is proportional to the number of collisions of molecules of both the gases taking place per second.

Consider a homogeneous gaseous reaction

$$A + B \rightarrow \text{Products}$$

Suppose a unit volume contains one molecule of each A and B, the probability of collision at any instant is 1 × 1 = 1.

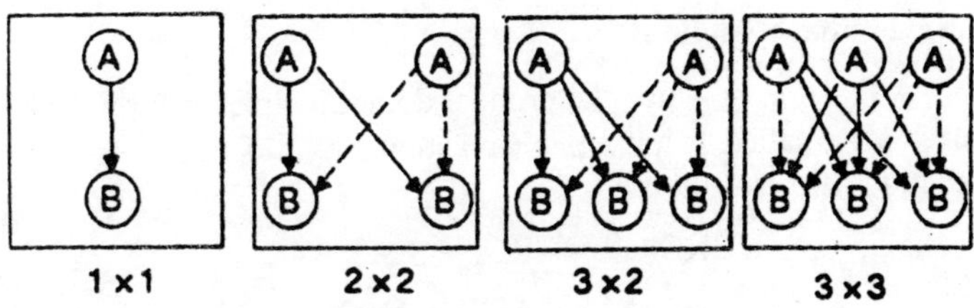

Fig. 4 : Collision of reacting molecules.

Where the number of molecules of A and B are two each then the possibility of collision is equal to $2 \times 2 = 4$. Suppose, there are two molecules of A and three molecules of B in it volume, the number of chances of collisions per unit time is $2 \times 3 = 6$. Similarly, if there are three molecules of each A and B per unit volume, the chances of collision will be $3 \times 3 = 9$.

Thus, chances of collisions are equal to the product of umber of molecules present per unit volume.

$\therefore$ Rate of reaction $\propto$ product of molecules of A and B per unit volume.

$$\propto [A][B]$$

This is exactly in accordance with the statement of the law of mass action.

Reversible Reactions

A reaction is said to be reversible if the composition of the reaction mixture on the approach of equilibrium at a given temperature is the same irrespective of the initial state of the system, *i.e.*, irrespective of the fact whether we start with the reactants or the products. A reversible reaction never tends to completion.

Some examples of reversible reactions are as follows:

$$Fe + 4H_2O \rightleftarrows Fe_3O_4 + 4H_2$$

$$N_2 + O_2 \rightleftarrows 2NO$$

$$H_2 + I_2 \rightleftarrows 2HI$$

The extent of reversibility of a reaction is governed by the nature of the reaction and the experimental conditions of temperature, concentrations of substances and pressure on the system, if gases are involved. Most of the reaction that we know are reversible in nature.

Irreversible Reactions

A reaction in which the product do not interact to form the original reactants is called an irreversible reaction. Some examples of irreversible reactions are as follows:

$$2KClO_3 \rightarrow KCl + 3O_2$$

$$NaCl + AgNO_3 \rightarrow AgCl + NaNO_3$$

Irreversible reactions proceed to completion in one direction.

Irreversible reactions have their equilibrium positions almost on one extreme.

Most of the precipitation reactions are irreversible.

EXPERIMENTAL PROOF FOR DYAMICAL EQUILIBRIUM

Now, it has been possible to establish the dynamic nature of chemical equilibrium experimentally in several cases. This has been demonstrated in the case involving hydrogen-iodine-hydrogen iodide reaction.

$$H_2(g) + I_2(g) \rightleftarrows 2HI(g)$$

The equilibrium of this reaction is attained at 448°C. As soon as this equilibrium is attained, a radioactive isotope of iodine is introduced into the reaction mixture. After sometime, it will be found that HI contains radioactive iodine whereas the relative amounts of HI, H_2 and I_2 do not get changed. This reveals that although no change is taking place in the relative amounts of the reactants and products, yet chemical reaction is taking place from left to right. But there occurs no increase i the amount of Hi. Therefore, it means that the reaction is taking place from left to right as well at the same rate. Thus, it means that although the system has been in equilibrium, yet the two opposing reactions have been proceeding and the equilibrium conditions are being maintained by a dynamic balance between the two reactions.

LIMITATIONS OF THE EQUATION FOR CHEMICAL EQUILIBRIUM

The equation for representing chemical equilibrium falls to reveal how far the reaction would proceed from left to right or right to left before the equilibrium is attained. Suppose the following equilibrium is considered:

$$N_2O_4(g) \rightleftarrows 2NO_2(g).$$

The above equation does not reveal what fraction of N_2O_4 gets dissociated into NO_2. If the temperature of the reaction is raised by 10°C, there occurs a considerable increase in the dissociation of N_2O_4 but even then the equation for the reaction remains the same.

The equation used for representing the chemical equilibrium fails to reveal how long it takes for a reaction to attain equilibrium. This can be understood by considering the equilibrium of water vapour with hydrogen and oxygen

$$H_2O(g) \rightleftarrows H_2(g) + \frac{1}{2} O_2(g)$$

Water is a highly stable compound. Its degree of dissociation is appreciably small. Even at 2000°C, its dissociation is mainly 0.6%. Thus, the equilibrium is lying towards much to the left. However, it has been verified experimentally that a mixture of hydrogen and oxygen remains as such for a sufficient long time without undergoing reaction to form water and establish the above equilibrium. The equation for the equilibrium fails to reveal that it requires so long to the equilibrium.

If an electric arc is struck in the hydrogen-oxygen mixture, the reaction starts instantaneously and the equilibrium is attained within minute or so. The equation of the chemical equilibrium fails to reveal the rate at which a chemical system comes to equilibrium and the conditions under which the equilibrium can be attained rapidly.

LOW OF MASS ACTION

A qualitative relation connecting the concentrations of the reactants and products with the amount of chemical change produced known as the low of mass action was proposed in 1867 by Guldbeg and Waage. This law can be stated as follows:

"*The rate at which a substance reacts a proportional to the active mass and the rate of a chemical reaction is proportional to the product of active masses of the reactants.*"

By the term active mass, it means the *molecular concentration.* In gases and solutions, it means the umber of gram molecules (*i.e.*, moles) present per litre. In the case of solids or pure liquids the active mass may be regarded as constant, *i.e.*, for sake of simplicity, it is taken as unity.

Active mass of a substance is expressed by enclosing the symbol or the formula of a substance i square brackets *i.e.*, [A] or by the symbol C at the base of which the symbol of the substance is placed, *i.e.*, C_A

Let us consider the following reversible reaction taking place at constant temperature:

$$A + B \rightleftarrows C + D$$

According to the law of mass action, the velocity of forward reaction: v_1 is given by

$$v_1 \propto C_A + C_B$$

$$v_1 \propto k_1 C_A + C_B \qquad ...(1)$$

where C_A and C_B are the molecular concentrations of A and B and k_1 is a constant which may be called the *velocity constant of the forward reaction.* The concentrations of A and B go on decreasing with the progress of the reaction and hence the velocity of the forward reaction gradually decreases with time as shown graphically in Fig. 5.

Since the reaction is reversible, the products as soon as they accumulate, start reacting to reform the reactants. The velocity of backward reaction is given by

$$v_2 = k_2 + C_C + C_D \quad ...(2)$$

where C_C ad C_D are the molecular concentrations of C and D and k_2 is the velocity constant of the backward reaction. As the concentrations of the products gradually increase, the velocity of the backward reaction will also steadily increase, as shown in Fig. 5.

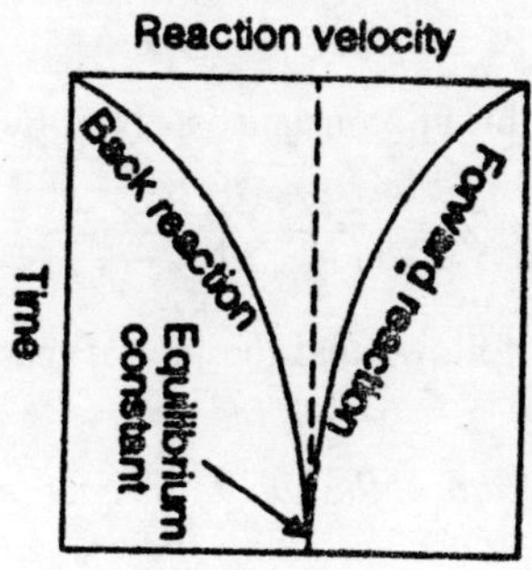

Fig. 5

Since the commencement of the reaction (zero time), the concentrations of the products gradually increase, the velocity of the backward reaction will also steadily increase. Ultimately a stage called *dynamic equilibrium* is reached when the velocity of toward reaction becomes equal to that of backward reaction, *i.e.*,

$$v_1 = v_2 \quad ...(3)$$

At this stage the number of products molecules produced per second in the forward reaction would be equal to the number of product molecules which would disappear in the backward reaction.

EQUILIBRIUM CONSTANT

On substituting Eqs. (1) and (2) in Eq. (3), we get

$$k_1 CAcB = k_2 C_C C_D$$

or $$\frac{k_1}{k_2} = K_C = \frac{C_C C_D}{C_A C_B} \qquad ...(4)$$

where K_C is a constant, called the equilibrium constants. It is defined as the ratio of velocity constants of two opposing reactions.

Eq. (4) is known as the equilibrium law equation.

When the reaction is of the type $2A \rightleftarrows C + D$, one may write

$$2A \rightleftarrows C + D$$

Hence the equilibrium constant is given by

$$k_C = \frac{C_C \times C_D}{C_A C_A} = \frac{C_C \times C_D}{C_A^2}$$

A general equation for a reversible reaction can be written as

$$aA + bB + ... \rightarrow cC + dD + ...$$

Hence the equilibrium constant is given by

$$k_C = \frac{C_C^c \times C_D^d \times ...}{C_A^a \times C_B^b \times ...} \qquad ...(5)$$

From Eq. (5), it follows that the law of mass action may be stated in another way

"*At a give temperature, the rate of a reaction is proportional to the molar concentration of arch reacting species raised to the power of the number of its molecules taking part in the reaction as represented in the stoichiometic equation describing the reaction.*"

In Eq. (5), when the concentrations are expressed in moles per litre, the equilibrium concentration is written as K_C.

In the case of gaseous reactions, the concentrations may be put proportional to the partial pressures. Hence the gaseous reactions, Eq. (5) may be put in the following form:

$$K_P = \frac{P_C^c \times P_C^d \times ...}{P_A^a \times P_B^b \times ...} \qquad ...(6)$$

where K_P is another equilibrium constant. The partial pressures are generally expressed in atmospheres.

Relationship between K_P and K_C. For an ideal gas,

$$P_1 V = nRT$$

or $$P_i = n\frac{RT}{V} = C_iRT \qquad ...(7)$$

where C is the concentration. On substituting the concentrations of various species from Eq. (7), in Eq. (6), we get

$$K_P = \frac{(C_CRT)^c \times (C_DRT)^d \times ...}{(C_ART)^a \times (C_BRT)^b \times ...}$$

$$K_p = \frac{C_C^c \times C_D^d \times ...}{C_A^a \times C_B^b \times ...} \times (RT)^{(c+d)-(a+b)}$$

$$K_p = K_C \times RT^{\Delta n}$$

where $$\Delta n = (c + d + ...) - (a + b + ...)$$

Equilibrium Constant : The equilibrium constant is defined as simply the ratio of the velocity constants of two opposing reactions.

OR

The equilibrium constant is the product of the concentrations of the reacting species, each concentration raised to the power which is the stoichiometric coefficient of the substance in the chemical equation.

In general, for the reaction represented by the equation,

$$aA + bB + cC \rightleftarrows lL + mM + nN$$

the equilibrium constant is give by

$$K_C = \frac{[K_1]}{[K_2]} = \frac{[L]^l[M]^m[N]^n}{[A]^a[B]^b[C]^c}$$

The equilibrium constant for the reaction $2SO_2 + O_2 \rightleftarrows 2SO_3$ is as follows.

$$K_c = \frac{[SO_3]^2}{[SO_2]^2[O_2]} = \frac{(\text{litre / mol})}{(\text{litre / mol})(\text{litre / mol})^2}$$

It means that the equilibrium constant for this reaction is constant.

FACTORS INFLUENCING EQUILIBRIUM CONSTANT

The various characteristics or the reactors that influence the equilibrium constants are as follows :

1. *Change in Temperature* : The equilibrium constant has a definite value for every chemical reaction at a given temperature. The value of equilibrium constant varies with temperature. It has also

the same value (at a give temperature) irrespective of the pressure on the system or the concentrations of the reactants and products.

2. *The mode of Expressing the Reaction :* If a reaction is represented in two opposite ways, the value of equilibrium constant in each case would be the reciprocal of the equilibrium constant i th other case For example,

$$A + B \rightleftarrows C + D, \quad K = \frac{C_c \times C_d}{C_A \times C_B}$$

$$C + D \rightleftarrows A + B, \quad K_1 = \frac{C_A \times C_B}{C_C \times C_D} = \frac{1}{K}$$

It means that if the equilibrium constant is large in one direction, it will be smaller in the other direction.

3. *Change of Concentration or Partial Pressure Units :* The units of equilibrium constants K_P and K_C are

$$K_P = (\text{Atm})^{\Delta n},$$

$$K_C = (\text{Mol dm}^{-3})^{\Delta n}$$

For reactions in which Δn is zero, *i.e.*, the number of moles of products is same as that of reactants, K has no units. For reactions in which Δn is not zero, any change in units expressing concentration or partial presents of the involved substances will alter the equilibrium constant.

4. *Representation of Stoichiometric Equation :* The value of equilibrium constant depends upon the stoichiometric equation representing the reversible reaction. If it is expressed in different ways, the value of K is affected. The formation of SO_3 from SO_2 and O_2 may be represented by either of the following stoichiometric equations.

$$2SO_2 + O_2 \rightleftarrows 2SO_3 \quad \text{...(i)}$$

$$SO_2 + \frac{1}{2} \rightleftarrows SO_3 \quad \text{...(ii)}$$

K_1 from reaction (i),

$$K_1 = \frac{C_{SO_3^2}}{C_{SO_2} \times C_{O2}} \quad \text{...(iii)}$$

K_2 from reaction (ii),

$$K_2 = \frac{C_{SO_3}}{C_{SO_2} \times C_{O2}^{1/2}} \quad \text{...(iv)}$$

From (iii) and (iv), we have

$$K_2 = \sqrt{K_1}$$

Significance of Equilibrium Constant

The equilibrium constant is regarded as a convenient measure of the extent to which a reaction goes before equilibrium is established. In order to illustrate the statement, we consider the following reaction:

$$2H_2O(g) \rightleftarrows 2H_2(g) + O_2(g);$$
$$K = 1.35 \times 10^{-11} \text{ mol l}^{-1} \text{ at } 1073 \text{ K}$$

As the value of K is very small, it means that there will be very small quantities of H_2 and O_2 before equilibrium is established. Thus, the reverse reaction is almost complete. Due to this reaction, a mixture of H_2 and O_2 is dangerously explosive.

Let us now consider another reaction

$$2O_3(g) \rightleftarrows 3O_2(g);$$
$$K = 10^{55} \text{ mol l}^{-1} \text{ at } 298 \text{ K}$$

because K is very large, the reaction mixture at equilibrium will contain a high proportion of oxygen and reaction will be nearly complete at equilibrium, *i.e.*, O_2 is more stable than O_3.

SOME SOLVED PROBLEMS

Problem 1:

In an experiment, at a total pressure of 10 atmospheres and 400°C, i the equilibrium mixture $2NH_3 \rightleftarrows N_2 + 3NH_2$ the ammonia was found to have dissociated to the extent of 96%. Calculate K_P for the reaction.

Solution:

For reaction:	NH_3	$\rightleftarrows$	N_2 +	$3H_2$
Initial	1 mole		0	0
At equilibrium	$(1 - x)$		$x/2$	$3x/2$
		$= 1 - x + x/2 + 3x/2 = 1 + x$		
Partial pressure	$\frac{(1-x)P}{(4-2x)}$		$\frac{x \,.\, P}{2(1+x)}$	$\frac{3x \,.\, P}{2(1+x)}$

We have $K_P = \frac{(p_{N_2})(p_{H_2})^3}{(pNH_3)}$

$$= \frac{\frac{xP}{1(1+x)}\left[\frac{3xP}{2(1+x)}\right]^3}{(1-x)/(1+x)P} = \frac{27x^4P^2}{16(1-x)^2(1+x)^2}$$

Substituting x = 0.96 and P = 10 atmosphere, we get

$$\therefore \qquad K_P = \frac{27(0.96)^4(10)^2}{16(1-0.96)^2(1+0.96)^2} = 23{,}320.$$

Problem 2:

Calculate K_P and K_C at 1 atmosphere and at 25°C for the reaction.

$$N_2O_4 = 2NO_2$$

$$(\Delta G^o = 1.1 \text{ K cal.})$$

Solution:

From equation, we have

$$\Delta G^o = -RT \ln K_P$$

Here $\Delta G^o = 1.1$ K cal = 1100 cals.

$$R = 2 \text{ cals.}$$

$$T = 253 + 25 = 298 \text{ K}$$

$$\therefore \qquad 1100 = -2 \times 298 \ln K_P$$

$$K_P = 0.1579 \text{ atm.}$$

The relation between K_P and K_C is

$$K_P = K_C(RT)^{\Delta n}$$

where $\Delta n = 2 - 1 = 1$

$$\therefore \qquad K_C = \frac{K_P}{RT}$$

Here $K_P = 0.1579$ atm., R = 0.082 litre-atm/degree

$$\therefore \qquad K_C = \frac{0.1579}{0.082 \times 298} = 6.461 \times 10^{-3} \text{ mole/litre}$$

Problem 3(a):

The equilibrium constant for this reaction $N_2 + 3H_2 \rightleftarrows 2NH_3$ at 400°C is 1.064×10^{-4} and at 500°C is 0.144×10^{-4}. Calculate the mean heat of formation of 1 mole of ammonia from its elements in the given range of temp.

Solution:

We have

$$\ln \frac{K_{C_2}}{K_{C_1}} = \frac{q_v}{R}\left(\frac{1}{T_1} - \frac{1}{T_2}\right)$$

$$2.303 \log \frac{K_{C_2}}{K_{C_1}} = \frac{q_v}{R}\left(\frac{T_2 - T_1}{T_1 T_2}\right)$$

or $$\log \frac{0.144 \times 10^{-4}}{1.64 \times 10^{-4}} = \frac{q_v}{1.987 \times 2.303} \times \left(\frac{773 - 673}{773 \times 673}\right)$$

or $$q_v = -25{,}150 \text{ cal.}$$

Thus the heat of reaction

$$N_2 + 3H_2 \rightleftarrows 2NH_3$$

is $$-25150 \text{ cal.}$$

In simple words, heat of formation of 2 moles of NH_3 is – 25,150. Therefore, heat of formation of mole of NH_3

$$= -\frac{1}{2} \times 25150 = -12575 \text{ cal.}$$

Problem 3(b):

Prove that

$$\left(\frac{\partial \ln K_P}{\partial T}\right) = \frac{\Delta H^\circ}{RT^2}$$

when $$\Delta G^\circ = -RT \ln K_P$$

Solution:

$$\Delta G^\circ = -RT \ln K_P \quad ...(1)$$

On differentiating with respect to temperature at constant pressure, we get

$$\left(\frac{\partial \Delta G^\circ}{\partial T}\right)_P = -R \ln K_P - RT\left(\frac{\partial \ln K_P}{\partial T}\right)_P$$

On multiplying both sides by T, we get

$$\left(\frac{\partial \Delta G^\circ}{\partial T}\right)_P = -RT \ln K_P - RT^2\left(\frac{\partial \ln K_P}{\partial T}\right)$$

$$= \Delta G^\circ - RT^2\left(\frac{\partial \ln K_P}{\partial T}\right)_P \quad \text{[use (1)] ...(2)]}$$

Gibbs-Helmholtz equation for substances in their standard states may be written as follows:

$$\Delta G^\circ = \Delta H^\circ + T\left(\frac{\partial \Delta G^\circ}{\partial T}\right)_P$$

$$T\left(\frac{\partial \Delta G^\circ}{\partial T}\right)_P = \Delta H^\circ - \Delta G^\circ \qquad ...(3)$$

From Eqs. (2) and (3), we have

$$\Delta H^\circ - \Delta G^\circ = -\Delta G^\circ + RT^2\left(\frac{\partial \ln K_P}{\partial T}\right)$$

$$-\frac{\Delta H^\circ}{RT^2} = \left(\frac{\partial \ln K_P}{\partial T}\right)$$

Problem 4:

The equilibrium constant of the reaction

$$2SO_2 + O_2 \rightleftarrows 2SO_3$$

at 528°C is 98.0 and at 680°C is 10.5 and find the heat of reaction.

Solution:

According to the integrated form of Van't Hoffs Isochore equation,

$$\log_{10} K_{p2} - \log_{10} K_{p1} = \frac{\Delta H}{2.303R}\left(\frac{1}{T_1} - \frac{1}{T_2}\right) = \frac{\Delta H}{2.303R}\left(\frac{T_2 - T_1}{T_1 T_2}\right)$$

where K_P is the equilibrium constant, ΔH is the heat of reaction.

$$K_{p2} = 10.5, \quad K_{p1} = 98.0$$

$$T_1 = 528 + 273 \rightleftharpoons 801 \text{ K},$$

$$T_2 = 680 + 283 = 953 \text{ K}$$

$$\log_{10} 10.5 - \log_{10} 98 = \frac{\Delta H}{2.303 \times 1.987}\left\{\frac{953 - 801}{953 \times 801}\right\}$$

$$(\log_{10} 10.5 - \log_{10} 98) = \left\{\frac{953 \times 801}{953 - 801}\right\} = \frac{\Delta H}{2.303 \times 1.987}$$

or

$$\Delta H = (2.303 \times 1.987)(\log_{10} 10.5 - \log_{10} 98)\left\{\frac{953 \times 801}{953 - 801}\right\}$$

$$= \frac{2.303 \times 1.987 \times (-0.97)801 \times 953}{152}$$

$$= -22290 \text{ cal.}$$

Problem 5(a):

The equilibrium constant K for the reaction

$$N_2 + 3H_2 \rightleftarrows 2NH_3$$

is 1.064 × 10^{-4} at 400°C and 0.144 × 10^{-4} at 500°C. Calculate the mean heat of formation of mole of ammonia from its elements in this temperature range,

Solution:

$$\log_{10} K_{p_2} - \log K_{p_1} = \frac{\Delta H}{2.303\, R}\left(\frac{T_2 - T_1}{T_1 T_2}\right)$$

$$\log_{10} 0.144 \times 10^{-4} - \log_{10} 1.64 \times 10^{4} = \frac{\Delta H}{2.303 \times 1.988}\left\{\frac{773 - 673}{773 \times 673}\right\}$$

as $\quad T_1 = 400 + 273 = 673$ K and $T_2 = 500 + 273 = 773$ K

or $\quad (\log_{10} 0.144 \times 10^{-4} - \log_{10} 1.64 \times 10^{-4})\dfrac{673 \times 773}{773 - 673} = \dfrac{\Delta H}{2.303 \times 1.987}$

or $\quad \Delta H = 2.303 \times 1.987 \times (\log_{10} 0.144 \times 10^{-4} - \log_{10} 1.64 \times 10^{-4} \times \dfrac{673 \times 773}{100}$

$$= \frac{2.303 \times 1.987 \times (-1.05648) \times 773 \times 673}{100} \text{ cal.}$$

$$\Delta H = -25150 \text{ cal.}$$

But this is the heat of reaction for the formation of 2 g moles of ammonia, hence heat at of formation of one g molecule.

$$= -\frac{25150}{2} = -12575 \text{ cal}$$

Negative sign shows that heat is evolved in this reaction.

Problem 5(b):

At 1000 k, water vapour at 1 atm has bee found to be dissociated into hydrogen and oxygen to the extent of 3 × 10^{-7}%. Calculate the free energy decrease of the system, assuming ideal behaviour.

Solution:

$$2H_2O \rightleftarrows 2H_2 + O_2$$

$$K_P = \frac{(p_{H_2}O)^2}{(p_{H_2})^2 (p_{O_2})}$$

If the pressure is 1 atm then

$$p_{H_2} = 3.0 \times 10^{-7},\ p_{O_2} = \frac{1}{2}(3.0 \times 10^{-7})$$

$$= 1.5 \times 10^{-7}$$

$$\because \quad K_P = \frac{1}{(3.0 \times 10^{-7})^2 (1.5 \times 10^{-7})}$$

$$= \frac{1}{1.35 \times 10^{-20}} = 7.4074 \times 10^{19}$$

$$\therefore \quad -\Delta G = RT \ln K_P = 2.303\ RT \log K_P$$

$$= 2.303 \times 8.314 \times 100 \times \log 7.4074 \times 10^{19}$$

$$= 380447.3 \text{ J} = 380.4473 \text{ kJ}$$

Problem 6:

At 375 K and total pressure of 1 atom sulphuryl chloride is dissociated according to the equation.

$$SO_2Cl_2 \rightleftarrows SO_2 + Cl_2$$

to the extent of 91.2%. Find K_P and the et work by this reaction at this temp.

Solution:

$$SO_2Cl_2 \rightleftarrows SO_2 + Cl_2$$

is $\quad K_P = \frac{\alpha^2 P}{1-\alpha}$

or $\quad K_P = \frac{\alpha^2 P}{(1-\alpha)(1+\alpha)}$

$$= \frac{(0.912)^2 \times 1}{(1-0.912)(1+0.912)} = 4.944$$

Net work w = et decrease in free energy = $-\lambda G$

But $\quad -\Delta G = RT \ln K_P$

$$= 2.303 \times 375 \times 8.314 \times \log 4.944$$

$$= 4983.6063 \text{ J}$$

Problem 7:

The equilibrium constant of the reaction

$$2SO_2 + O_2 \rightleftarrows 2SO_3$$

at 528°C is 98 and at 680°C is 10.5. Find the heat of reaction.

Solution:

According to the integrated form of Vant Hoff's Isochore equation.

$$\log_{10} K_{P_2} - \log_{10} K_{P_1} = \frac{\Delta H}{2.303}\left(\frac{1}{T_1} - \frac{1}{T_2}\right) = \frac{\Delta H}{2.303}\left(\frac{T_2 - T_1}{T_1 T_2}\right)$$

where K_P is the equilibrium constant, ΔH is the heat of reaction.

$$K_{P_1} = 10.5,\ K_{P_2} = 98.0$$

$$T_1 = 528 + 273 = 801 \text{ K},$$

$$T_2 = 680 + 273 = 953 \text{ K}$$

$$\therefore \log_{10}10.5 - \log_{10}98 = \frac{\Delta H}{2.303 \times 8.314}\left\{\frac{953 - 801}{953 \times 801}\right\}$$

$$(\log_{10}10.5 - \log_{10}98) \times \left\{\frac{953 \times 801}{953 - 801}\right\} = \frac{\Delta H}{2.303 \times 8.314}$$

$$\Delta H = (2.303 \times 8.314)(\log_{10} 98 - \log_{10} 10.5)$$

$$\left\{\frac{953 \times 801}{953 - 801}\right\} \text{J}$$

$$= \frac{2.303 \times 8.314 \times (-0.97) \times 953 \times 801}{152} \text{ J}$$

$$\Delta H = -93273.332 \text{ J.}$$

Problem 8(a):

The temperature dependence of the equilibrium constant of the isomerization reaction D $\rightleftarrows$ is give by the relation

$$K_P = 4.814 - \frac{2059 \text{ K}}{T}$$

where T is in kelvin. Calculate the entropy of isomerization at 25°C and ΔG° and ΔH° at the same temperature.

Solution:

At 25°C (= 298 K), we will have

In $$K_P = 4.814 - \frac{2059\ K}{298\ K} = -2.095$$

Now $$\Delta G^\circ = -RT \ln K_P$$

$$= -(8.314\ JK^{-1}\ mol^{-1})(298\ K)(-2.095)$$

$$= 5190.5\ J\ mol^{-1}$$

Now the differentiation of ln K_P would give the expression

$$d \ln K_P = -(2059\ K)d(1/T)$$

or $$\frac{d \ln K_P}{d(1/T)} = -(2059\ K)$$

Hence $$-\frac{\Delta H^\circ}{R} = -(2059\ K)$$

or $$\Delta H^\circ = (2059\ K)(8.314\ J\ K^{-1}\ mol^{-1})$$

$$= 17118.5\ J\ mol^{-1}$$

Finally $$\Delta S^\circ = \frac{\Delta H^\circ - \Delta G^\circ}{T}$$

$$= \frac{(17118.5 - 5190.5)\ J\ mol^{-1}}{298\ K}$$

$$= 40.03\ J\ K^{-1}\ mol^{-1}$$

Problem 8(b):

At 250°C, $PCl_5(g)$ dissociates according to

$$PCl_5(g) \rightleftarrows PCl_3(g) + Cl_2(g)$$

to the extent of 80% under a total pressure of 1 atm. Find (i) K_P (ii) the total volume of the system and (iii) the concentration of each component in mol/litre if 1 mol of PCl_5 was taken initially.

Solution:

At equilibrium we will have

$$PCl_5(g) \rightleftarrows PCl_3(g) + Cl_2(g)$$

1 mol – 0.8 mol   0.8 mol   0.8 mol

Total amount of species = 1.8 mol.

Partial pressure of $PCl_5 = \frac{0.2}{1.8} \times 1\ atm$

Partial pressure of $PCl_3 = \frac{0.8}{1.8} \times 1$ atm

Partial pressure of $Cl_2 = \frac{0.8}{1.8} \times 1$ atm

Now $$K_P = \frac{p_{PCl_3} p_{Cl_2}}{p_{PCl_5}} = \frac{(0.8 \text{ atm}/1.8)^2}{(0.2 \text{ atm}/1.8)} = 1.778 \text{ atm}$$

(ii) The total volume of the system is given as

$$V = \frac{nRT}{p}$$

$$= \frac{(1.8 \text{ mol})(0.082 \text{ atm LK}^{-1} \text{ mol}^{-1})(523 \text{ K})}{(1.8 \text{ mol})}$$

$$= 42.89 \text{ L}$$

(iii) Concentrations of the species are as follows:

$$[PCl_5] = \frac{0.2 \text{ mol}}{42.89 \text{ L}}$$

$$= 4.663 \times 10^{-3} \text{ mol L}^{-1}$$

$$[PCl_3] = [Cl_2] = \frac{1.8 \text{ mol}}{42.89 \text{ L}}$$

$$= 4.20 \times 10^{-2} \text{ mol L}^{-1}$$

Problem 8(c):

The equilibrium constant for the reaction $H_2 + I_2 \rightleftarrows 2HI$ at 450°C is found to be 64. If 6 moles of hydrogen are mixed with 3 moles of iodine ion a one-litre vessel at this temperature, what will be the concentration of each of three components when the equilibrium is attained?

Solution:

For reaction:	H_2 +	$I_2 \rightleftarrows$*	$2HI$
Initial	6 moles	3 moles	0
At equilibrium	(6 – x) moles	(3 – x) moles	2x moles
Concentration	(6 – x)/1 moles litre^{-1}	(3 – x)/1 moles litre^{-1}	2x/1 moles litre^{-1}

$\therefore$ $$K_C = \frac{[HI]^2}{[H_2][I_2]}$$

$$= \frac{(2x)^2}{(6-x)(3-x)} = \frac{4x^2}{18-9x+x^2} = 64$$

or $$\frac{x^2}{18-9x+x^2} = 16$$

or $$x^2 = 16 \times 18 - 16 \times 9x + 16x^2$$

or $$15x^2 - 144x + 288 = 0$$

or $$5x^2 - 48x + 96 = 0$$

$$x = \frac{48 \pm \sqrt{(48)^2 - 4 \times 5 \times 96}}{2 \times 15}$$

$$= \frac{48 \pm \sqrt{384}}{30} = \frac{48 \pm 19.75}{30} = 2.258 \text{ or } 0.942$$

Hence, composition of equilibrium mixture will be

H_2 = 6 – 2.258 (0.942) = 3.742 or 5.058 mole/litre

I_2 = 3 – 2.258 (0.942) = 3.742 or 2.058 mole/litre

HI = 2 × 2.258 (0.942) = 4.516 or 1.884 mole/litre.

Problem 9:

When ethyl alcohol and acetic acid are mixed i equivalent proportions equilibrium is attained when 2/3rds of the acid and alcohol are consumed. How much ester will be present in the equilibrium mixture if we start with 240 g of acetic acid and 138 g of alcohol?

Solution:

For reaction:	CH_3COOH +	C_2H_5OH ⇄	$CH_3COOC_2H_5$	$+H_2O$
Initial	a moles	b moles	0	0
At equilibrium	(a – x) moles	(b – x) moles	x mole	x mole
Concentration	(a – x)/V	(b – x)/V	x/V	x/V

We have $K_C = \frac{[x/V] \times [x/V]}{[(a-x)/V][(b-x)/V]} = \frac{x^2}{(a-x)(b-x)}$

In first case if a = 1, b = 1, x = 2/3

$$K_C = \frac{(2/3)^2}{\left(1-\frac{1}{3}\right)\left(1-\frac{2}{3}\right)} = \frac{[2/3] \times [2/3]}{[1/3] \times [1/3]} = 4.0$$

In second case a = 240 g/60 = 4 moles (∵ mole wt of acid = 6)

and b = 138 g/46 = 3 mole (∵ mole wt. of alcohol = 46)

$$\therefore \quad K_C = 4 = \frac{x^2}{(4-x)(3-x)}$$

or $$x^2 = 4(12 - 7x + x^2) = 48 - 28x + 4x^2$$

or $$3x^2 - 28x + 36 = 0$$

$$\therefore \quad x = \frac{28 \pm \sqrt{(28)^2 - 4 \times 3 \times 36}}{2 \times 3} = 2.26$$

as the second value is not admissible. Hence, the amount of ester present at equilibrium

$$= 2.26 \text{ mole} = 2.26 \times 88 = 198.88 \text{ g.}$$

Problem 10:

Phosphorus pentachloride is 82% dissociated into phosphorus trichloride and chlorine, all gaseous, at 250°C and 1 atmosphere. Calculate the equilibrium constant of dissociation

Solution:

For reaction:	PCl_5	PCl_3 +	Cl_2
Initial	1 moles	0	0
At equilibrium	(1 − x) mole	x mole	x mole
		= 1 − x + x + x = 1 + x mole	
Partial pressure	$\frac{(1-x)}{(1+x)}P$	$\frac{x.P}{(1+x)}$	$\frac{x.P}{(1+x)}$

$$\text{We have } K_P = \frac{p_{Cl_2} \times p_{PCl_3}}{p_{PCl_5}} = \frac{\frac{x.P}{1+x} \times \frac{x.P}{1+x}}{\frac{(1-x)P}{(1+x)}} = \frac{x^2P}{(1-x^2)}$$

Substituting x = 0.82; P = 1 atmosphere, we et

$$\therefore \quad K_P = \frac{(0.82)^2 \times 1}{(1+0.82)(1-0.82)}$$

$$= \frac{0.82 \times 0.82}{1.82 \times 0.18}$$

$$= 2.577 \text{ atmosphere.}$$

Problem 11:

The equilibrium constant for the reaction $N_2O_4 \rightleftarrows 2NO_2$ at 500°C is found to be 640. If the pressure of the gas mixture is 180 mm, calculate the percentage dissociation of N_2O_4. At what pressure will it be half dissociated?

Solution:

For reaction: $N_2O_4 \rightleftarrows 2NO_2$

Initial 1 moles 0

At equilibrium(1 – x) mole 2x mole = 1 – x + 2x = (1 + x) mole

Partial pressure $\frac{(1-x)}{(1+x)}P$ $\frac{2xP}{(1+x)}$

We have
$$K_P = \frac{(p_{NO_2})^2}{p_{N_2O_4}} = \frac{\left[\frac{2x\,.\,P}{(1+x)}\right]^2}{\frac{(1-x)P}{(1+x)}} = \frac{4x^2P}{(1-x^2)}$$

Substituting K_P = 640; P = 180/760 = 9/38 atmospheres, we get

$$\therefore \quad 640 = \frac{4x^2 \times 9}{(1-x^2) \times 38}$$

or $$36x^2 = 640 \times 38 - 640 \times 38x^2$$

or $$24{,}356x^2 = 24{,}320$$

or $$x = 0.9992$$ or 99.92% dissociated at 500°C.

Let N_2O_4 is half-dissociated at P, *i.e.*, x = 0.5

$$640 = K_P = \frac{4x^2P}{(1-x^2)} = \frac{4 \times (0.5)^2 \times P}{(1+0.5)(1-0.5)} = \frac{P}{0.75}$$

$$P = 640 \times 0.75$$
$$= 480 \text{ atmosphere.}$$

Problem 12:

When a mixture of nitrogen and hydrogen in the volume ration of 1 : 3 allowed to attain equilibrium at 400°C and 1000 atm. 0.2491 mole fraction of ammonia is formed. Calculate the equilibrium constant for the above reaction at the experimental temperature and pressure.

Solution:

For reaction:	N_2	+	$3H_2$	$\rightleftarrows$	$2NH_3$
Initial	1 mole		3 moles		0
At equilibrium	$(1 - x)$		$(3 - 3x)$		$2x$

$$= 1 - x + 3 - 3x + 2x = 4 - 2x \text{ mole}$$

Partial pressure $\quad \dfrac{(1-x)P}{(4-2x)} \qquad \dfrac{(3-3x)P}{(4-2x)} \qquad \dfrac{2xP}{(4-2x)}$

We have:
$$K_P = \frac{(p_{NH_3})^2}{p_{N_2} \times (p_{H_2})^3}$$

$$= \frac{\left[\dfrac{2xP}{(4-2x)}\right]^2}{\left[\dfrac{(1-x)P}{(4-2x)}\right]\left[\dfrac{(3-3x)P}{(4-2x)}\right]^3} = \frac{16x^2(2-x)^2}{27P^2(1-x)^4}$$

Here P = 1000 atmospheres and mole fraction of NH_3 = 0.2491 = $2x(4 - 2x)$, we get

$$\therefore \quad 2x = 0.9964 - 0.4982x$$

or
$$2.4982x = 0.9964$$

or
$$x = 0.9964/2.4982 = 0.4$$

then,
$$K_P = \frac{16 \times (0.4)^2 \times (1.6)^2}{27 \times (1000)^2 \times (0.6)^2}$$

$$= 1.617 \times 10^{-6} \text{ atmospheres}^{-2}$$

2

MEASUREMENT OF LOWERING OF VAPOUR PRESSURE

INTRODUCTION

Raoult measured the individual vapour pressures of a liquid and the solution by this method. He introduced the liquid or the solution into Toricellia vacuum of a barometer tube and measured the depression of the mercury level. The method is neither practicable nor accurate as lowering of vapour pressure has been too small.

MANOMETRIC METHOD

The vapour pressure of a liquid or solution can be conveniently determined with the help of a manometer. The bulb B is charged with the liquid or solution. The air in the connecting tube is then removed with a vacuum pup. If the stopcock is closed, the pressure inside is due only to the vapour evaporating from the solution or liquid. This method is generally used for aqueous solutions. The manometric liquid can be mercury or n-butyl phthalate which has low density and low volatility.

WALKER AND OSTWALD'S METHOD

(i) A current of dry air is passed through two sets of bulbs A ad B containing the solution and the solvent respectively and placed in a constant temperature bath.

(ii) As the air passes through the solution it takes up with it the vapours of the solvent. The amount of vapour take up by the air is proportional to the vapour pressure p_s of the solution. When the air enters the bulbs B it takes up more solvent vapours for the vapour pressure p of the solvent is more than that of solution. The amount of vapour take fro bulbs B is proportional to difference in the vapour pressure of solvent and the solution $(p - p_s)$. When the air comes out of the bulbs B, it is saturated with solvent and the amount of solvent present in it is proportional to p, the vapour pressure of the solvent.

(iii) The amount of solvent carried by air from the bulbs A and B can be known from the loss in weight of bulbs A and B respectively.

$\therefore$ Loss in weight of A is $\propto p_s$

Loss in weight of B is $\propto p - p_s$

Loss in weight of A + B is $\propto p$

$$\text{Hence } \frac{\text{Loss in wt. of B}}{\text{total loss in wt. of A + B}} = \frac{p - p_s}{p}$$

(iv) The total loss in weight of bulbs A and B can also be determined by finding out the increase in weight of $CaCl_2$ tube attached at the end of bulbs B. In that case:

$$\frac{\text{Loss in wt. of B}}{\text{Gain in wt. of } CaCl_2 \text{ tube}} = \frac{p - p_s}{p}$$

Thus the relative lowering of vapour pressure can be calculated from the values of different weights and consequently molecular weight of solute is calculated.

Osmosis

When a solution is separated from its solvent by a semipermeable membrane, the solvent molecules pass through it into solution to have uniform concentration on both sides of the membrane.

"*The spontaneous flow of solvent into a solution or from a ore dilute to a concentrated solution through a semipermeable membrane is known as osmosis.*"

Illustration of Osmosis

Take two eggs of equal size whose outer shells have been removed. Put one of them in distilled water and the other in saturated salt solution. After few hours the egg placed in water swells up while the one in salt solution shrinks. It is due to the fact that in case of egg placed distilled water, the water enters the concentrated egg fluid while in other case it goes out of the egg into more concentrated salt solution to have uniform concentration in and out.

Difference from Diffusion

(i) In osmosis the presence of semipermeable membrane is essential.

(ii) In osmosis it is only the solvent molecules which flow out whist in diffusion both the solute as well as solvent molecules flow out in opposite directions.

In osmosis the flow of solvent molecules start from the solution of lower concentration to that of higher concentration while in diffusion the molecules over form a region of higher concentration to those of lower concentration.

Osmosis differs from diffusion in the following respects:

Table 1

Osmosis	*Diffusion*
1. There is a flow of solvenet into the solution through a semipermeable membrane.	1. There is a flow of both the solvenet and the solute and no semipermeable membrane is required.
2. Solvent flows from the solution of lower concentration to solution of higher concentration.	2. Solution flows from higher concentration to lower concentration until an equilibrium in concentration is achieved.

Semipermeable Membrane

A membrane which allows the solvent but to the solute molecules to pass through it is known as semi-permeable membrane.

Problems:

(i) The membrane surrounding the plant and animal cells are almost semipermeable. In fact pig bladder has bee used as semi-permeable membrane in the past in different experiments.

(ii) A layer of phenol saturated with water acts as a semipermeable layer for a solution of calcium nitrate as the latter is not soluble in phenol.

(iii) The gelatinous precipitates of inorganic substance such as calcium phosphate and copper ferrocyanide are now commonly used as semipermeable membranes. Of laboratory experiments copper ferrocyanide is referred and is constructed by depositing it in the porous walls of a battery pot.

Preparation

The pronous pot is first thoroughly cleaned by washing it with acid, water, alkali and finally with distilled water. The air bubbles enclosed in the pores are removed by forcing water through the pot under pressure. The pot is then filled with a 2.5% solution of copper sulphate and placed in a vessel containing solution of potassium ferocynide of same strength. A bottomless platinum cylinder is placed surrounding the pot and acts as anode.

On passing electric current, copper ions go out of the pot while ferocynide ions come into the pot. They meet some where in the walls of the pot to form a compact and stout membrane of copper ferocyanide.

$$2CuSO_4 + K_4Fe(CN)_6 \rightarrow \underset{\text{Copper ferrocyanide}}{Cu_2Fe(CN)_6} + 2K_2SO_4$$

When the membrane is being formed, the electrical resistance of the cell increase and finally reaches a maximum value. On completion of membrane, the current stops and the bell ceases ringing.

SOME THEORIES OF OSMOTIC PRESSURE AND SUPERMEABILITY

Though we know that osmotic pressure is bought into existence only when the solution is separated fro the solvent by a semipermeable membrane, we still do not know how a semipermeable membrane acts.

1. Sieve Theory

Traube first suggested that the semipermeable membrane acts like a molecular seive whereby the molecules bigger than the solvent are retarded, while the solvent molecules filter through. Thus the solvent flows from a region of higher solvent concentration to one of lower concentration. But this theory raised eyebrows when it was seen that solute molecules even smaller than the resolvent were retarded by the membrane.

Later, when the theory of chemical interaction at the poles due to the polar groups of the protein molecules lining the pores became known, this theory of *molecular seiving* became popular. Bigelow assumes that the membrane acts like a set of very fine capillary tubes. The walls of these capillaries are neither wetted by water nor by solution. Thus each capillary will have at its opposite ends water ad solution separated by a small gap. The vapour pressure other side will decide the flow of solvent molecules. As vapour pressure is greater on the water side, transfer of water will take place from solvent into solution. This theory has a general acceptance.

2. Chemical Theory

Amstrong proposed a chemical theory whereby the solvent was supposed to form a kind of loose chemical compound with the membrane, and was later spilt off again. But due to lack of experimental evidence this theory was rejected.

3. Preferential Solubility Theory

Another theory which speaks of preferential solubility was well received. It is supported by experimental proof and it states that solvent dissolves in the membrane, diffuse through ad is give off at the other end. When a mixture of nitrogen and hydrogen gas is passed through a heated palladium box, hydrogen is give off at the other end. This separation is achieved because H_2 dissolves in palladium, diffuses through ad is give off at the other end.

4. Kinetic Theory

The kinetic theory suggests that the osmotic pressure is due to bombardment of the molecules of solvent o the membrane. As the solution side carries solute molecules too, the bombardment per unit area of surface will be less. Hence the solvent molecules will diffuse slowly from this side than the solvent side resulting in an attempt to make the number equal.

5. Hydrostatic Theory

The hydrostatic theory suggests that the entrance of solvent into the solution is supposed to be due to the existence of an attractive force of the solvent for the solute. This might be due to variation in surface tension.

6. Vapour Pressure Theory

According to this view the supermeable membrane is a vapour sieve *i.e.*, the vapours of the solvent can only pass through it. As the vapour pressure of the pure solvent is higher than that of the solution, the solvent molecules can pass through the membrane to the solution side. The theory satisfactorily explains the phenomenon of osmosis.

7. Solution Theory

According to this theory put forward by Therrmite, membrane is permeable to those substances which dissolve in it and is impermeable to those substances which are insoluble in it. For example calcium nitrate is insoluble in phenol but is soluble in water and hence a layer of phenol slipped between calcium nitrate and water acts as semi-preamble membrane. This theory is widely accepted.

WHY EXACTLY DOES OSMOSIS TAKE PLACE?

The answer becomes very simple if we look at it thermodynamically. When a solute is dissolved in a solvent, the energy of the solvent molecules is reduced considerably because of the solute-solvent intractions. When this solution is separated from the pure solvent by a seipermeable membrane rapid movement of the solvent molecules across the barrier occurs in the direction of the solution. The hydrostatic pressure so developed ion the copartner containing the solution is enough to increase the free energy of the solvent molecules in the solution. When the free energy of the solvent molecules is the solution is restored to a value equal to that of the pure solvent, an equilibrium is achieved and osmosis stops. More the user of solute particles in the solution, greater is the osmotic pressure developed. Thus, osmotic pressure is one of the four colligative properties that a solution possesses (vapour pressure, rise in boiling point, depression of freezing point, being the others). Under ideal conditions, it is just like gas pressure, directly proportional to the absolute temperature and to concentration, and is independent of the chemical nature of the dissolved material.

Osmotic Pressure

When a solution is separated fro the pure solvent by a semipermeable membrane, a diffusion of solvent takes place through the membrane from the pure solvent into the solution. In case, if two solutions of different concentrations are separated by semipermeable membrane, the solvent molecules will move from the dilute solution to the ore concentrated solution side. *This spontaneous movement of solvent molecules through a seipermeable membrane from a region of dilute solution into a region of concentrated solution is termed as osmosis.*

If several solutes are present osmosis will take place if the semipermeable membrane is impermeable to any one of the various kinds of molecules or ions. In order to demonstrate osmosis, a thistle funel, whose lower end is tied with a animal membrane, is filled with a sugar solution and the kept in a beaker which contains distilled water. Due to osmosis, water will flow into the funel through the membrane. A considerable rise in level of the liquid in the thistle funel is noticed. As the liquid rises in the tube, a hydrostatic pressure is developed.

This pressure increases the tendency for the solvent molecules to over from the solution back into the pure solvent and operates against the force causing osmosis. As a result, an equilibrium condition is finally reached in which the hydrostatic pressure is sufficient to prevent further diffusion. *That hydrostatic pressure acting on the solution side of the semipermeable membrane that is just sufficient to prevent osmosis is termed as the osmotic pressure of the solution.*

If a liquid rises to a height h,

Osmotic pressure = hdg where d is the density of the solution.

Another definition of osmotic pressure : Suppose a solution is kept in a cup whose walls contain a seipermeable membrane. This cup is tightly closed by a rubber piston. The, this cup is kept in a bigger beaker containing water (solvent).

Due to osmosis, the water will move from the beaker it of the cup through the semipermeable membrane. This tendency can be overcome by applying pressure on the solution by keeping increasing weights on the piston. As soon as a proper weight is placed, no osmosis takes place. *This excess pressure acting on the solution which prevents osmosis is known as osmotic pressure of the solution.*

Another definition of osmotic pressure : Another definition of the osmotic pressure may be given in terms of vapour pressure. We know that the vapour pressure of a solution is lower than that of its solvent. Let us place two beakers. one filled with a solution and other with solvent under reduced pressure. The vapour phase acts as a semipermeable membrane. The vapours of solvent from the beaker containing solvent will travel towards the beaker containing solution. The movement of vapour can be checked by applying external pressure o the solution so that the vapour pressure of the solution just becomes equal to that of solvent. Thus, the osmotic pressure may be defined as *the external pressure which must be applied o the solution in order to increase its vapour pressure to such a extent that it becomes equal to the vapour pressure of the solvent.*

VAN'T HOFF'S THEORY

From measurements of osmotic pressure Van Hoff observed that certain laws which are obeyed by gases are also obeyed by dilute solutions. The Boyle's law, Charle's Law, Gas equation and Avogadro's hypothesis are also applicable to dilute solutions and are shown below:

(i) *Boyle Vant Hoff Law :* Pfeffer's results indicate that the osmotic pressure of solution is directly proportional to its concentration.

$$\therefore \quad P \propto C \propto \frac{1}{V},$$

(where P, osmotic pressure; C, concentration in moles per litre; V, volume of solution containing 1 mole of solute)

$$PV = \text{constant (at constant T)}$$

I other words "*At constant temperature, the product of osmotic pressure and volume is constant.*"

This is exactly similar to Boyle's Law for gases.

(ii) *Pressure Temperature Law :* The results of osmotic pressure measurement indicate that if the concentration of solution is kept the same, the osmotic pressure varies directly as absolute temperature. The concentration we know is inversely proportional to volume and since the volume changes in liquids with rise of temperature are negligible, hence P μ T at constant volume.

In other words "*At constant volume, the osmotic pressure of a dilute solution is directly proportional to absolute temperature.*"

This is similar to Charle's law for gases.

(iii) *General equation for solutions :* From the above two laws:

$$P \propto \frac{1}{V} \text{ and also } P \propto T,$$

$$\therefore \quad P \propto \frac{T}{V} \quad \text{or} \quad PV \propto T$$

Hence PV = ST, (where S is solution constant).

This equation is exactly similar to general gas equation and the value of S (.0821 litre-atoms) corresponds about exactly with R obtained for gases.

(iv) *Avogadro-Van Hoff Law :* For a given solution

$$P_1V_1 = ST_1$$

(where V_1 is the volume containing 1 mole of the solute).

For another solution,

$$P_2V_2 = ST_2$$

(where V_2 is the volume containing 1 mole of solute).

Now if $P_1 = P_2$ and $T_1 = T_2$; then $V_1 = V_2$.

In other words, *equal volume of solutions exerting same osmotic pressure at same temperature contain equal number of solute molecules."*

This is similar to Avogadro's law of gases.

Van Hoff, on the basis of analogous behaviour of dilute solutions to gases as discussed above, put forward in 1885 the theory of dilute solutions called Van Hoff's theory of dilute solution. According to it:

"*A substance in solution behaves exactly like a gas and the osmotic pressure of a dilute solution is equal to the pressure which the solute would exert if it were a gas at the same temperature and occupying the same volume as the solution*".

In other words the theory may be stated as "The solute molecules in a dilute solution play the same role as is played by the gas molecules in a gas, the osmotic pressure being analogue of gas pressure."

MEASUREMENT OF OSMOTIC PRESSURE

Different methods are used in measuring the osmotic pressure:

(i) Pfeffer's method,

(ii) Morse and Frazer's method

(iii) Berkley and Hartley's method

(iv) Electronic Osmometre

But the one mostly employed in measuring the osmotic pressure is due to Berkely and Hartley.

1. Pfeffer's Method

Pfeffer's method (1817) : Pfeffer gave a direct method to measure osmotic pressure of solution. Here a porous pot (A) containing membrane of copper ferrocyanide in its wall is cemented to a glass tube (B) containing solution and attached to manometer (M). The porous pot is kept in pure solvent. The osmotic pressure exerted by the solution is given by the manometer. The method is of historical interest only and has been given up.

Disadvantages:

(i) The osmotic pressure developed in dilute solution is very great and bursts the semipermeable membrane used.

(ii) It takes a long time to register the final osmotic pressure.

2. Morse and Frazer's Method

An improvement of this method was made by H.N. Morse and J.C.C.W. Frazer ad their collaborators (1901-1923). They prepared sophisticated semi-permeable membranes of copper ferrocyanide by using an electric current. This made the membranes more tough ad they were capable of withstanding high pressures without leakage. Chamber I which has the semipermeable membrane deposited on its walls is filled with water. Tube 2 is also filled with water. Chamber 3 is filled with solution. When osmosis begins, solvent flows from 1 to 3 increasing the hydrostatic pressure in vessel 3. This hydrostatic pressure developed, which is equal to osmotic pressure, is measured by a manometer attached to 4.

3. Berkley and Hartley's Method (1909)

This is the most simple, rapid and accurate method of measuring osmotic pressure. It is based on the fact that counter pressure applied o the solution so as to prevent osmosis is a measure of osmotic pressure.

(i) The apparatus consists of two concentric tubes, the inner one being that of porcelain having electrically deposited semipermeable membrane of copper ferrocyanide in it's walls. The two ends of the inner tube are connected to a capillary T on one side and dropping funnel D on the other. The outer tube is made of gun metal and is fitted with an arrangement for applying definite pressure.

(ii) In the angular space between the two tubes is introduced the solution whose osmotic pressure is to be measured while the inner tube is filled with water by means of dropping funnel upto a definite level T in the capillary tube.

(iii) Due to osmosis the water from the inner tube tends to pass into the solution which is indicated by the downward motion of the water meniscus in the capillary tube T. The external pressure is applied on the solution to prevent the flow of water and consequently lowering of meniscus. This pressure so applied is equal to osmotic pressure and is directly measured by means of pressure gauge attached to a piston.

Advantage

(i) The equilibrium is established ore quickly and hence *time required* for measurement of osmotic pressure is much smaller.

(ii) The strain on membrane is much less and hence there is *no chance* of its *bursting* out as in other methods.

(iii) The strength of solution does not change during measurements and hence is more accurate.

For solutions in non-aqueous solvents where suitable seipermeable membranes are available only with difficulty another method `called Towned's porous disc method is quite useful. For biological fluid's however, De Varies plasmolytic method (1884) is very convenient.

4. Electronic Osmometer

Sorensen in 1917 first published the data of his measurements of osmotic pressure o protein solutions under carefully controlled conditions. His results and those of Starlin's on serum which was reproduced using collodion membranes and better designed modern osmometers, showed that a strict control of pH and salt concentration, and knowledge of the final concentration of protein are mandatory while determining the osmotic pressures of protein solutions. The osmometers which were used by Sorensen and others in the first half of 20th century though efficient, took several days for standardization. The osmotic event was rapid but the rise due to capillarity in the manometer tube required several days to reach an equilibrium value. The osmometers designed later were easy to handle and more accurate.

The manometers in these latter type of instruments contained organic liquids, paraffin, toluene or alcohols with which equilibrium could be attained in a very short time. The sample requirement also was low. But these instruments required a rigorous thermostatic control within 0.004°. Later, this problem was circumvented by the invention of an electronic osmometer by Rowe and Abrams in 1957. This could be operated without thermostatic control and the sample for determination was needed in a very small volume (only 0.5 ml). A detailed discussion of the apparatus is give as follows;

An osmometer adapted for rapid measurements of osmotic pressure using small volumes of sample was designed as above by Rowe and Abrams in 1957. The U-tube has a semi-permeable membrane arranged in its arm B which holds the protein solution (sample) above it while the solvent lies beneath it. Arm 'A' also contains the same solvent. In the horizontal portion of the tube lies a very thin platinum foil which is connected further to a mechano-electronic transducer in circuit with a galvanometer. A controlled air pressure is continuously exerted on the protein solution.

When the tap is closed, the system if not in equilibrium, tries to attain it. The solvent passes through the membrane and the resulting pressure change displaces the platinum foil. This is indicated on the galvanometer scale. If the system is in equilibrium, no deflection can be observed in the galvanometer. The osmotic pressure exerted by the solution can be calculated from the formula,

$$\text{O.P.} = P + H_p r_p - H_s \rho_s - H_t$$

where P = pressure applied to protein solution,

H_p = height of column of protein above membrane,

H_s = height of column of solvent above membrane,

H_t = effect due to difference of surface tension between protein solution and solvent,

ρ_p = density of protein solution, and

ρ_s = density of solvent.

MOLECULAR WEIGHT FROM OSMOTIC PRESSURE

According to van Hoff, dilute solutions obey the general gas equation $PV = ST$

where P = Osmotic pressure

V = Volume in litres containing 1 mole

S = Solution constant, .0821 litre-atm

T = Absolute temperature.

Thus from this equation

$$V = \frac{ST}{P} \qquad \text{...(9)}$$

Let w g of solute be dissolved in litre of solution for determining osmotic pressure then the amount which will be present in V litres must be 1 mole (*i.e.,* mol wt. in g.)

or w g is present in 1 litre.

$\therefore$ M, the mol. wt. in g will be in $= \dfrac{M}{w}$ litres

This must be the volume $V = \dfrac{M}{w}$ ...(ii)

From equations (i) and (iii), we get

$$\frac{M}{w} = \frac{ST}{P} \quad \text{or} \quad M = \frac{S \times T \times w}{P}$$

or
$$M = \frac{.0821 \times T \times w}{P}$$

Thus, knowing w, T and P for the give solution, M, the mol. mass of substance can be calculated.

This method is, however, confronted with *difficulty* if the substance whose molecular weight is to be determined is an electrolyte. In such cases as the molecules dissociate the values of osmotic pressure actually observed and calculated from the solution equation (PV = ST) are not similar and hence the molecular weight calculated from osmotic pressure is not correct.

Isotonic Solutions

The solutions having same osmotic pressure are know as *isotonic solutions*. When such solutions are separated by a semipermeable membrane, no osmosis takes place.

We know for two solutions, PV = P′V′.

If the solutions are isotonic P = P′, then V must be same as V′. Hence the volumes of these solutions containing 1 mole of the respective substances must be same.

In other words: "*The isotonic solutions have same molar concentration.*"

For example : 5% solution of urea is isotonic with 15% grape sugar solution as the osmotic pressure exerted by these solution is same. It is due to the fact that their molar concentrations are equal (*i.e.,* $\frac{5}{60}$ and $\frac{15}{180}$).

The knowledge of isotonic solutions is very helpful in determining molecular weight. Thus if the molecular weight of one of the substance forming isotonic solutions be known that of the other can be calculated.

If an unknown solution, has lower osmotic pressure than the give known solution, the former is called *hypotonic solution but* in case the unknown solution has higher osmotic pressure it is called *hypertonic solution.*

Abnormal Osmotic Pressure

When the osmotic pressure of electrolytes is measured, the values observed are much higher than those calculated from the value of V and T in the equation PV = ST. It is due to the fact that in such compounds the molecules dissociate or associate to the two or more ions and since the osmotic pressure of the solution depends upon the number of particles present in the give volume, the values observed will be higher in case of electrolytes. The abnormal values of osmotic pressure observed in case of electrolytes (*i.e.,* different from those calculated) is known as *abnormal osmotic pressure.* These abnormal values of osmotic pressure in case of electrolytes are useful in determining the degree of dissociation of the electrolytes.

Plasolysis

The protoplasmic layer lining the cell walls of plant cells or of red blood cells is seipermeable; being permeable to water but is impermeable to the substances dissolved in the cellular fluid. This cellular fluid has an osmotic pressure of its own. Thus when a plant cell is placed a solution of lower osmotic pressure (Hypotonic) than that of the cell fluid, water penetrates into it and is consequently swells up. But if on the other hand, the plant cell is placed in a solution of greater osmotic pressure (Hypertonic) water diffuse out of the cell fluid resulting in contraction or even partial collapse of the membrane. The phenomenon is known as *plasmolysis.* This phenomenon has proved to be very useful to the measurement of relative osmotic pressure, De Vries (1884) developed the method. The cells are placed in the solution of different concentrations and the structures of the cells are observed under a microscope. The solution will have the same osmotic pressure (Isotonic) as that of cellular fluid in which plasolysis does not take place.

ELEVATION OF BOILING POINT

The boiling point of a liquid is the temperature at which its vapour pressure becomes equal to the atmospheric pressure. But the vapour pressure of a solution is always lowered due to the addition of non-volatile solute. Therefore, the solution has to be heated to a higher temperature so that its vapour pressure becomes equal to the atmospheric pressure. Hence, the boiling point of solution having a non-volatile solute is higher than the boiling point of the pure solvent In other words, the boiling point of the solution is said to be elevated. *The difference between the boiling points of solution and pure solvent at a certain constant pressure is known as the elevation of boiling point of the solution.*

In order to understand the relation between elevation of boiling point and lowering of vapour pressure, the vapour pressure curves are considered. The vapour pressures of solvent and two solutions I and II have been plotted against temperature. The boiling points of solvent and two solutions are represented by T, T_1 and T_2 respectively.

For dilute solutions, the curves are considered to be parallel straight lines. From the similar triangles CDB and AEC, we get,

$$\frac{AC}{BC} = \frac{EC}{CD}$$

or $$\frac{p - p_2}{p - p_1} = \frac{T_2 - T}{T_1 - T} \quad \text{or} \quad \frac{\Delta p_2}{\Delta p_1} = \frac{\Delta T_2}{\Delta T_1} \qquad ...(1)$$

In the above expressions, p, p_1 and p_2 denote the vapour pressures of solvent, solution I and solution II.

From equation (1), it is clear that

Lowering of vapour pressure ∝ Elevation of boiling point

or $$\Delta p \propto \Delta T_b$$

or $$\Delta T_b \propto \Delta p$$

But according to Raoult's law, we have

$$\Delta p \propto x_B$$

$$\therefore \quad \Delta T_b \propto x_B \quad \text{or} \quad \Delta T = K x_B \qquad ...(2)$$

where K is the proportionality constant and x_B is the mole fraction of solute which is defined as follows:

$$x_B = \frac{\frac{w_B}{M_B}}{\frac{w_A}{M_A} + \frac{w_B}{M_B}} \qquad ...(3)$$

where w_A and w_B denote the masses of solvent A and solute B respectively;

M_A and M_B are the molecular asses of solute and solvent respectively. If a solution is dilute, $\frac{w_B}{M_B} << \frac{w_A}{M_A}$; then equation (3) becomes as follows:

$$x_B = \frac{\frac{w_B}{M_B}}{\frac{w_A}{M_A}} = \frac{M_A w_B}{M_B w_A} = n_B \frac{M_A}{w_A} \quad ...(4)$$

From the definition of molality, n_A/w_B represents the molality m of the solution if w_A is taken in kilogram units. Hence equation (4) becomes as

$$x_B = m \,.\, M_A \quad ...(5)$$

On substituting equation (5) in (2), we get

$$\Delta T_b = K \,.\, m \,.\, M_A = k_b \,.\, m \quad ...(6)$$

where k_b (= K . M_A) is a new constant called molal boiling point elevation constant for the solvent, when *m* = 1 molal, equation (6) becomes as $\Delta T_b = k_b$.

Thus, *the molal boiling point elevation constant for the solvent is numerically equal to the elevation in boiling point which is observed for 1 molal solution.*

From the definition of molality, we have

$$m = \frac{1000\, w_B}{M_B w_A} \quad ...(7)$$

On substituting equation (7) in (6), we get

$$\Delta T_b = k_b \,.\, \frac{1000\, w_B}{M_B w_A}$$

or

$$M_B = \frac{1000\, k_b}{\Delta T_b} \,.\, \frac{w_B}{w_A} \quad ...(8)$$

Equation (8) is of much use because this may be used to find out molecular masses (M_B) of solutes provided the remaining quantities k_b, ΔT_b, w_A and w_B are known.

Units of k_b : The values of k_b are expressed in degrees/molality or as K/m or °C/m.

Molal elevation boiling point constants have characteristic values for different solvents (Table 2).

Table 2 : Molal Boiling Point Elevation Constants for Some Solvents

S.No.	*Solvent*	*b.p. (K)*	k_b*(K/m)*
(i)	H_2O	373.0	0.52
(ii)	C_2H_5OH	351.5	1.20
(iii)	C_6H_6	353.5	2.53
(iv)	$CHCl_3$	334.4	3.63
(v)	CCl_4	350.0	5.03
(vi)	CS_2	319.4	2.34
(vii)	$C_4H_{10}O$(Ether)	307.8	2.02

MEASUREMENT OF BOILING POINT ELEVATION

There are several methods available for the measurement of the elevation of boiling point. Some of these are outlined below.

1. Landsberger-Walker method

This method was introduced by Landsberger and modified by Walker.

Apparatus : The apparatus used in this method is consists of :

(i) An inner tube with a hole in its side and graduated in ml;

(ii) A boiling flask which sends solvent vapour into the graduated tube through a 'rosehead' (a bulb with several holes);

(iii) An outer tube which receives hot solvent vapour issuing from the side-hole of the inner tube;

(iv) A thermometer reading to 0.01 K, dipping in solvent or solution in the inner tube.

Procedure : Pure solvent is kept in the graduated tube and vapour of the same solvent boiling in a separation flask is passed into it. The vapour causes the solvent in the tube to boil by its latent heat of condensation. When the solvent starts boiling and temperature becomes constant, its boiling point is recorded.

Now the supply of vapour is temporarily cut off and a weighed pellet of the solute is dropped into the solvent in the inner tube. The solvent vapour is again allowed to pass through until the boiling point of the solution is reached and this recorded. The solvent vapour is then cut off, thermometer and rosehead raised out of the solution, and the volume of the solution read. From a difference in the boiling points of solvent and solution, the molecular weight of the solute can be found out by using the expression

$$m = \frac{1000 \times K_b \times w}{\Delta T \times W}$$

where w = weight of solute taken, W = weight of solvent which is given by the volume of solvent (or solution) measured in l multiplied by the density of the solvent at its boiling point.

2. Cottrell's Method

A method better than Landsberger-Walker method was devised by Cottrell (1910).

Apparatus : It consists of : (i) a graduated boiling tube containing solvent or solution; (ii) a reflux condense which returns the vaporised solvent to the boiling tube; (iii) a thermometer reading to 0.01 K, enclosed in a glass hood; (iv) A small inverted funnel with a narrow stem which branches into three jets projecting at the thermometer bulb.

Beckmann Thermometer: It is a differential thermometer. It is designed to measure small charges in temperature and to the temperature itself. It has a large bulb at the bottom of a fine capillary tube. The scale is calibrated from 0 to 6 K and subdivided into 0.01 K. The unique feature of this thermometer, however, is the small reservoir of mercury at the top. The amount of mercury in this reservoir can be decreased or increased by tapping the thermometer gently. In this way the thermometer is adjusted so that the level of mercury thread will rest any desired point o the scale when the instrument is placed in the boiling (or freezing) solvent.

Procedure : The apparatus is fitted up solvent is kept in the boiling tube with a porcelain piece lying in it. It is heated o a small flame (micro burer). As the solution starts boiling, solvent vapour arising from the porcelain piece pumps the boiling liquid into the narrow set.

Thus a mixture of solvent vapour and boiling liquid is continuously sprayed around the thermometer bulb. The temperature soon becomes constant and the boiling point of the pure solvent is recorded.

Now a weighed pellet of the solute is added to the solvent and the boiling point of the solution noted as the temperature becomes steady. Also, the volume of the solution in the boiling tube can be noted. The difference of the boiling temperature of the solvent and solute gives the elevation of boiling point. While calculating the molecular weight of solute the volume of solution is converted into mass by multiplying with density of solvent at its boiling point.

DEPRESSION IN FREEZING POINT

Freezing point is the temperature at which solid and liquid states of a substance have the same vapour pressure. But the vapour pressure of solution is less than that of the solvent. Therefore, the freezing point of the solution will be lower than that of its solvent. The difference between the freezing points of the pure solvent and solution is called the depression in freezing point of the solution. In order to understand the depression in freezing point, the curves between vapour pressure and temperature for solid solvent, liquid solvent and two solutions of different concentrations are plotted.

T_f is the temperature at which the curve AC due to solid solvent intersects with the curve AE due to liquid solvent. Therefore, T_f is the freezing point of the solvent. Similarly, T_1 and T_2 are the freezing points of two solutions I and II respectively.

For very dilute solutions, the curves are considered to be straight lines. From similar triangles ABC and DCE, we get

$$\frac{BC}{CE} = \frac{AB}{ED}$$

or $$\frac{p - p_2}{p - p_1} = \frac{T - T_2}{T - T_1}$$

or $$\frac{\Delta p_2}{\Delta p_1} = \frac{\Delta T_2}{\Delta T_1} \quad ...(1)$$

From the above relation, it follows that the lowering of vapour pressure is directly proportional to the depression of freezing point *i.e.,*

$$\Delta p \propto T_f \quad \text{or} \quad \Delta T_f \propto \Delta p \quad ...(2)$$

From Raoult's law, we have

$$\Delta p \propto x_B$$

$$\therefore \quad \Delta T_f \propto x_B \quad \text{or} \quad \Delta T_f = Kx_B \quad ...(3)$$

where K is a proportionality constant. The mole fraction x_B of solute is defined mathematically as:

$$x_B = \frac{\frac{w_B}{m_B}}{\frac{w_A}{M_A} + \frac{w_B}{M_B}} \qquad ...(4)$$

where w_A = mass of solvent, w_B = mass of solute, and

M_A = molecular mass of solvent and

M_B = molecular mass of solute.

For dilute solutions, $\frac{w_B}{M_B} << \frac{w_A}{M_A}$, equation (4) becomes as follows:

$$x_B = \frac{\frac{w_B}{M_B}}{\frac{w_A}{M_A}} = \frac{M_A}{M_B} \times \frac{w_B}{w_A} = n_B \frac{M_A}{w_A} \qquad ...(5)$$

In the above equation, n_B/w_A represents the molality of the solution if w_A, mass of the solvent, is taken in kilogram units. Therefore, equation (5) becomes as follows:

$$x_B = m \,.\, M_A \qquad ...(6)$$

On substituting the above equation in equation (3), we get

$$\Delta T_f = K \,.\, M_A \,.\, m = k_f m \qquad ...(7)$$

where k_f (KM_A) is a ew constant called *molal freezing point depression constant* which relates the depression in freezing point for the solution with its molality.

In order to define k_f, put m = 1 in equation (7), we get

$$\Delta T_f = k_f$$

From the above relation, it follows that "*Molal freezing point depression constant for the solvent is numerically equal to the depression in freezing point for 1 molal solution.*"

The molality of a solution is defined mathematically as

$$m = \frac{1000\, w_B}{M_B w_A} \qquad ...(8)$$

On substituting equation (8) in (7), we get

$$\Delta T_f = k_f \frac{1000\, w_B}{M_B w_A} \qquad ...(9)$$

or $$M_B = \frac{1000\, k_f \cdot w_B}{\Delta T_f \cdot w_A} \qquad ...(10)$$

Equation (10) is of a great practical importance because it permits the calculation of molecular mass of solute (M_B) provided the remaining quantities k_f, w_B, ΔT_f and w_A are known.

Unit of K_f: It will be degree/molality or K/m or °C/m

Values of k_f for some solvents are given in Table 3.

Table 3 : Freezing Point Depression Constants for Some Solvents

S.No.	*Solvent*	*f.p. (K)*	*k_f(K/m)*
(i)	H_2O	273.0	1.86
(ii)	C_2H_5OH	155.7	1.99
(iii)	C_6H_6	278.6	5.12
(iv)	$CHCl_3$	209.6	4.70
(v)	CCl_4	250.5	31.8
(vi)	CS_2	164.2	3.83
(vii)	$C_4H_{10}O$ (Ether)	156.9	1.76

Measurement of Freezing Point Depression

Two methods given below are used:

1. Beckmann's Method

This is the cost *widely used method* and gives sufficient accurate results.

(i) The apparatus consists of a big tube A having a side tube for introducing the solute.

(ii) It is fitted up with stirrer and Beckmann's Thermometer. The latter can read even $\frac{1}{100}$th of a degree.

(iii) The tube is filled with a definite weight of the solvent, sufficient to dip the thermometer bulb and is suspended in another tube B wide enough to provide sufficient air space between them.

(iv) This whose system is placed in an outer vessel containing freezing mixture whose temperature is about 5° below the freezing point of the pure solvent.

(v) The space between the two test tubes serves as an jacket and prevents direct heat conduction and consequent super cooling.

(vi) On placing the tubes in the freezing mixture the temperature of the solvent in the inner tube gradually falls and due to *super cooling* even goes below the true freezing point of the solvent. On rapid stirring, however, the solid begins to separate and the mercury rises rapidly in the thermometer due to latent heat set free and finally become steady. The temperature is noted which is the freezing point of the pure solvent.

(vii) The inner tube is taken out to remelt the solvent and a known weight of solute is introduced into it. The tube is the placed back in its former place and the freezing point of the solution is determined as before. The difference in the two readings gives the depression of the freezing point. By substituting the value in the equation, the molecular weight of the solute can be calculated.

2. Limitations of the Method

(i) Since the relation between molecular weight and depression of freezing point is derived from Raoult's law which is applicable to dilute solution only, *the concentration of solute in solution must be very low* otherwise abnormal values are obtained.

(ii) Since the depression of the freezing point is proportional to number of molecules of the solute, the method gives *abnormal value for substances which associate or dissociate* in solution. The observed molecular weight corresponds to true mol. wt. in case of compounds which either dissociate nor associate *i.e.*, non-electrolytes.

(iii) The method is not *applicable to volatile substance* for part of it may vaporize and due to which the concentration of the solution may chance and abnormal values may be obtained.

(iv) The substance should be perfectly soluble in solvent chosen and must remain in liquid phase during the experiment. The method thus *cannot be applied to insoluble or sparingly soluble substance.*

Calculation of Molecular Weight

Applying the expression

$$m = \frac{1000 \times K_f \times w}{\Delta T \times W}$$

$$\text{We have } m = \frac{1000 \times 3.86 \times 0.124}{0.324 \times 25} = 59.83$$

Thus the molecular weight (or relative molecular mass) of X is 59.83.

RAST'S CAMPHOR METHOD

This method due to Rast (1922) is used for determination of molecular weights of solutes which are soluble in molten comphor. The freezing point depressions are so large that an ordinary thermometer can be used.

Pure camphor is powered and introduced into a capillary tube which is sealed at the upper end. This is tied along a thermometer and heated in a glyceron bath. The melting point of campho is recorded. The a weighted amount of solute and camphor (about 10 times as much) are melted in test tube with the open end sealed. The solution of solute in camphor is cooled in air. After solidification, the mixture is powdered and introduced into a capillary tube which is sealed. Its melting point is recorded as before. The difference of the melting point of pure camphor and the mixture, gives the depression of freezing point. I modern practice, electrical heating apparatus is used for a quick determination of melting points of campho as also the mixture.

The molal depression constant of pure campho is 40°C. But since the laboratory campho may not be very pure, it is necessary to find the depression constant for the particular sample of camphor used by a preliminary experiment with a solute of known molecular weight.

ABNORMAL COLLIGATIVE PROPERTIES OF SOLUTIONS

in the derivation of colligative properties, it was assumed that the molecular form of the solute remains unchanged in solution. Also, the solutions are dilute and behave ideally. In such cases, experimental value of the colligative property is in agreement with the theoretically calculated value. However, there are certain substances like solutions of salts, acids or bases in water or acetic acid in benzene where the experimental value differs considerably from the calculated value. Such solutions are said to be *abnormal solutions*. The abnormalities observed in such solutions are of two types:

(i) association of the solute molecules, and

(ii) dissociation of solute molecules.

Association leads to a decease in the number of solute particles and hence the colligative properties will show lower values. In case of dissociation, the number of solute particles increases ad consequently, the colligative properties will show abnormally enhanced values.

In order to account for the abnormal behaviour of such solutions, van't Hoff's introduced a factor "i" which is called the *van't Hoff factor* and is defined as the ratio of the *experimental value of a colligative property to the calculated value of that property, i.e.,*

$$i = \frac{\text{Experimental value of the colligative property}}{\text{Calculated value of the property when the solution behaves ideally}}$$

Since the colligative property is proportional to number of solute particles in solution, hence

$$i = \frac{\text{Actual number of particles present in solution}}{\text{Number of particles in solution if it behaves ideally}}$$

or, we may write

$$i = \frac{(\Delta T_b)_{obs}}{(\Delta T_b)_{cal}} = \frac{(\Delta T_f)_{obs}}{(\Delta T_f)_{cal}} = \frac{(\Delta P / P^0)_{obs}}{(\Delta P / P^0)_{cal}} = \frac{\pi_{obs}}{\pi_{cal}} = \frac{M_{cal}}{M_{obs}} \quad ...(1)$$

when M is the molar mass of the solute and ΔT_b, ΔT_f, $\Delta P/P^0$ and π are the boiling point elevation, freezing point depression, relative lowering of vapour pressure and the osmotic pressure of the solution respectively. The subscripts 'obs' and 'cal' refer to the experimental and calculated values of the colligative properties.

(i) *Dissociation of solute :* Consider an electrolyte A_xB_y which partly dissociates in solution yielding x ions of A^{y+} ad y ions of B^{x-} and if α is the degree of dissociation, *i.e.,* the fraction of the total umber of molecules which dissociates and C the initial concentration of the solute, then the dissociation equilibrium in solution can be represented as

	$A_xB_y \rightleftharpoons$	xA^{y+} +	yB^{x-}
Initial concentration	C	0	0
Concentration at equilibrium	$C(1-\alpha)$	$Cx\alpha$	$Cy\alpha$

The total number of moles at equilibrium

$$= Cx\alpha + Cy\alpha + C(1-\alpha)$$

$$= C[1 + \alpha + x\alpha + y\alpha]$$

$$= C[1 + \alpha(x + y - 1)]$$

Hence $$i = \frac{C[1 + \alpha(x + y - 1)]}{C}$$

or the degree of dissociation α is given as follows:

$$\alpha = \frac{i - 1}{(x + y - 1)} \quad ...(2)$$

Equation (2) is applicable to any colligative property and provides an important method for calculating the degree of dissociation of a solute. If $\alpha = 1$, *i.e.*, the dissociation is complete, $i = x + y$, the observed colligative property will be $x + y$ times the calculated value.

On the other hand, when no dissociation occurs, $\alpha = 0$ and $i = 1$, the calculated and observed values will be equal.

(ii) *Association of solute*: Consider the association of a solute A into its associated from $(A)_n$ according to the reaction.

$$n_A = (A)_n$$

where n is the number of molecules of solute which combine to form an associated species. If C is the initial concentration and α the degree of association of the solute, at the equilibrium the number of moles of the undissociated solute is $C(1 - \alpha)$ and that of associated form is $\frac{C\alpha}{n}$. The total number of moles in solution is given by

$$C(1 - \alpha) + \frac{C\alpha}{n}$$

or $$C\left(1 - \alpha + \frac{\alpha}{n}\right)$$

Hence the van't Hoff factor,

$$i = \frac{C\left(1 - \alpha + \frac{\alpha}{n}\right)}{C} = 1 - \alpha + \frac{\alpha}{n} = \left[1 + \left(\frac{1}{n} - 1\right)\alpha\right]$$

or $$\alpha = \frac{i - 1}{\frac{1}{n} - 1} \quad ...(3)$$

If association is complete, *i.e.*, $\alpha = 1$, $i = 1/n$, the observed value of a colligative property is $1/n$ times the calculated value and if $\alpha = 0$, no association occurs in solution, *i.e.*, $i = 1$ and the observed and calculated values will be equal.

Colligative Properties

Those properties of the solutions which depend only upon the total number of molecules of the solute per unit volume and not on its chemical nature are called colligative properties. Four main colligative properties of solutions are as follows:

(i) Lowering of vapour pressure,

(ii) Elevation of boiling point,

(iii) Depression in freezing point, and

(iv) Osmotic pressure.

The colligative properties are widely used for the determination of molecular masses for substances.

MATHEMATICAL EXPRESSION

From the above definition of colligative properties, it follows that if two solutions are made from different components, they may show identical values of colligative properties which are dependent only one mole fractions in the solution. Thus,

Colligative property measured $\propto$ Mole fraction of the solute

Suppose a system of two components A and B (A is solvent and B is the solute) is considered. In this system, w_A is the molecular mass of the solvent whose molecular ass is M_A. Similarly, in this solution, w_B is the molecular mass of the solute whose molecular mass is M_B.

Now number of moles of solvent,

$$n_A = \frac{w_A}{M_A}$$

and number of moles of solute,

$$n_B = \frac{w_B}{M_B}$$

Hence, mole fraction of solvent,

$$x_A = \frac{n_A}{n_A + n_B}$$

and mole fraction of solute,

$$x_B = \frac{n_B}{n_A + n_B}$$

But according to the definition,

$$\text{Colligative property} \propto \frac{n_B}{n_A + n_B}$$

or $$\text{Colligative property} = \gamma \frac{n_B}{n_A + n_B} = \gamma \frac{\frac{w_B}{M_B}}{\frac{w_A}{M_A} + \frac{w_B}{M_A}} \quad ...(1)$$

In the above equation, γ is proportionality constant which depends upon the nature of colligative property.

Equation (1) is very useful because it can be used for calculating the value of any of the involved factors provided the values of all the rest are known.

RELATIVE LOWERING OF VAPOUR PRESSURE

The vapour pressure of a liquid is lowered when a non-volatile solute is dissolved in it and the lowering is proportional to the amount of solute dissolved.

Suppose the pure liquid has vapour pressure p and the solution has vapour pressure p_s, the lowering will be $p - p_s$.

"*The ratio of the lowering of vapour pressure to the vapour pressure of the pure solvent* $\frac{p - p_s}{p}$ *is known as relative lowering of vapour pressure.*"

Though the vapour pressure of a liquid increases rapidly with temperature, the relative lowering of vapour pressure is constant at all temperatures for a given dilute solution.

Raoult's Law

It states that: "*The relative lowering of vapour pressure is equal to the ratio of the number of molecules of the solute and the total number of molecules in the solution i.e., the molar fraction of the solute present in solution.*"

Mathematically it can be represented as:

$$\frac{p - p_s}{p} = \frac{n}{n + N}$$

where n and N are the number of moles of solute and the solvent respectively.

Suppose number of moles of solute = n

Suppose number of moles of solvent = N

$\therefore$ Number of g moles of solute in solution = $\dfrac{n}{n + N}$

Thus fraction of moles of solute in solution = $\dfrac{N}{n + N}$

Hence vapour pressure of solution will be proportional to both of them

i.e., $$p_s \propto \frac{n}{n + N} \quad \text{and} \quad p_s \propto \frac{N}{n + N}$$

But as the solution is very dilute n is very small, the value of $\dfrac{n}{n + N}$ becomes very small. So that fist proportionality can be neglected and hence we have only

$$p_s \propto \frac{N}{n + N}$$

or $$p_s = k\frac{N}{n + N} \text{ where k is constant.} \quad ...(i)$$

Transforming (i) into an equation of pure solvent we put p in place of p_s and n = 0 and we get:

$$p = k \quad ...(ii)$$

Substituting the value of k from (ii) in (i), we get:

$$p_s = p \times \frac{N}{n + N}$$

or $$\frac{p_s}{p} = \frac{N}{n + N}$$

Subtracting both the sides from 1 we get,

$$1 - \frac{p_s}{p} = 1 - \frac{N}{n + N}$$

or $$\frac{p - p_s}{p} = \frac{n + N - N}{n +} = \frac{n}{n + N}$$

or $$\frac{p - p_s}{p} = \frac{n}{n + N} \quad ...(iii)$$

This is a Raoult's Law.

Applicability of Raoult's Law

The Raoult's law equation derived as above is very useful in determination of molecular mass of solute. It can be written as:

$$\frac{p - p_s}{p} = \frac{w/m}{\frac{w}{m} + \frac{W}{M}}$$

Thus knowing other terms in the equation, m, the molecular mass of solute can be calculated.

Thus the equation can be utilized in calculation the molecular mass of dissolved substance when the relative lowering of vapour pressure produced by a known weight of the solute in a known weight of solvent is known.

RELATION BETWEEN LOWERING OF VAPOUR PRESSURE AND OSMOTIC PRESSURE

Consider a tube the lower part of which is fitted with a semipermeable membrane and containing the solution. It is immersed in a vessel containing the pure solvent and the whole arrangement is placed in larger outer vessel which is evacuated. The osmosis take place and the level of solution in the tube rises till the hydrostatic pressure developed o the solution prevents osmosis. The osmotic pressure of the solution is thus

$$P = h \times \rho \times g \qquad ...(1)$$

where P is osmotic pressure, h the height to which the solution rises in the where tube; ρ the density of the solvent for being *dilute* solution; g is the Acceleration due to gravity. Now the pressure of vapours at the level of surface A of the solution must be same inside and outside the tube. If p is the vapour pressure of pure solvent and p_s vapour pressure of the solvent over solution, the difference between them is equal to the pressure of the vapour column of height h. This is

$$p - p_s = p \times g \times d ...(ii)$$

(where d is the density of the vapour)

Let M be the molecular ass of the solvent vapour, then the volume V occupied by Mg of vapours of pressure p and temperature, T is given by the gas equation:

$$pV = RT \quad \text{or} \quad V = \frac{RT}{p}$$

$\therefore$ d, the density of vapour $= \dfrac{M}{V} = \dfrac{M}{RT/p} = \dfrac{pM}{RT}$

or $$d = \frac{pM}{RT} \quad ...(iii)$$

Substituting the value of h and d from (i) and (iii) in (ii) we get;

$$p - p_s = \frac{p}{\rho g} \times g \times \frac{pM}{RT}$$

or $$\frac{p - p_s}{p} = \frac{M}{RT\rho} \times P \quad ...(iv)$$

or $$p - p_s = \frac{Mp}{RT\rho} \times P$$

Hence the factor $\dfrac{Mp}{RT\rho}$ is constant for a solvent of molecular mass M, density ρ, vapour pressure p at temperature T.

$$p - p_s \propto P$$

Hence lowering of vapour pressure is directly proportional to osmotic pressure of solution.

Derivation of Raoult's Law Equation

Let n moles of solute be dissolved in volume V of the solution then according to V and Hoff's equation of solution:

$$PV = nRT$$

or $$P = \frac{nRT}{V} \quad ...(v)$$

Substituting the value of P from (v) in (iv) we get;

$$\frac{p - p_s}{p} = \frac{M}{R\rho T} \times \frac{nST}{V} = \frac{Mn}{\rho V}$$

But ρV is product of density and volume = Mass of solution, F

$$\therefore \quad \frac{p - p_s}{p} = \frac{Mn}{F} = \frac{n}{F/M} = \frac{n}{N}$$

where F/M is the number of moles of solvent in ass F of the solvent.

$$\therefore \quad \frac{p - p_s}{p} = \frac{n}{N} \quad ...(vi)$$

This is modified form of Raoult's law equation.

SOME SOLVED PROBLEMS

Problem 1:

The vapour pressure of water at 293 K is 17.540 mm Hg and the vapour pressure of a solution of 0.10824 kg. of a nonvolatile solute in 1 kg of water at the same temperature is 17.354 mm Hg. Calculate the molar mass of the solute.

Solution:

Here

$$p = 17.540 \text{ mm Hg}$$

$$= \frac{1.013 \times 10^5 \times 17.54}{760} \text{ Nm}^{-2} = 2.337 \times 10^3 \text{ N m}^{-2}$$

$$p_s = 17.354 \text{ mm} = \frac{1.013 \times 10^5 \times 17.354}{760} \text{ Nm}^{-2}$$

$$= 2.313 \times 10^3 \text{ N m}^{-2}$$

$$w = 0.10824 \text{ kg}$$

$$m = ? \quad W = 1 \text{ kg} \quad M = 18 \times 10^{-3} \text{ kg mol}^{-1}$$

We know $$\frac{p - p_s}{p} = \frac{wM}{mW}$$

or $$m = \frac{wM}{W} \times \frac{p}{p - p_s}$$

$$= \frac{(0.10824 \text{ kg})(18 \times 10^{-3} \text{ kg mol}^{-1})(2.337 \times 10^3 \text{ N n}^{-2})}{(1 \text{ kg})(0.024 \times 10^3 \text{ N m}^{-2})}$$

$$= 0.1837 \text{ kg mol}^{-1}$$

Problem 2(a):

The vapour pressure of ether (mol wt = 74) is 442 m Hg at 293 K. If 3 g of a compound A are dissolved in 50 g of ether at this temperature, the vapour pressure falls to 426 mm Hg. Calculate the molecular mass of A. Assume that the solution of A in ether is very dilute.

Solution:

Here the approximate form of the Raoult's law Equation will be used.

$$\frac{p - p_s}{p} = \frac{w}{\frac{m}{W / M}} = \frac{wM}{mW} \qquad \text{...(i)}$$

In this case:

w, the weight of solute (A) = 3 g

W, the weight of solvent (ether) = 50 g

m, the mol mass of A = ?

M, the mol wt of solvent (ether) = 74

p, the vapour pressure of solvent (ether) = 442 mm

p_s, the vapour pressure of solution = 426 m

Substituting the values in equation (1),

$$\frac{426 - 442}{426} = \frac{3 \times 74}{m \times 50}$$

whence $m = 123$

Thus the molecular mass of A is 123.

Problem 2(b):

The boiling point of a solution containing 0.20 g of a substance X in 20.00 g of ether is 0.17 K higher than that of pure ether. Calculate the molecular weight of X. Boiling point constant of ether per 1 kg is 2.16 K.

Solution:

Applying the expression

$$m = \frac{100 \times K_b \times w}{\Delta T \times W}$$

In this case, we have

$\Delta T = 0.17$ K

$K_b = 2.16$

$w = 0.20$ g

$W = 20.00$ g

Substituting values

$$m = \frac{1000 \times 2.16 \times 0.20}{0.17 \times 20.00}$$

$$m = 127.81.$$

Problem 3:

The boiling point of chloroform was raised by 0.325 K when 5.141 × 10⁻⁴ kg of anthracene was dissolved in 35 × 10⁻³ kg of chloroform. Calculate the molar mass of the solute (molal elevation constant for chloroform is 3.9).

Solution:

Here, $\Delta T_b = 0.325$ K $\quad W = 35 \times 10^{-3}$ kg

$K_b = 3.9$ K kg mol^{-1} $\quad w = 5.141 \times 10^{-4}$ kg

$m = ?$

We know,

$$\Delta T_b = K_b \frac{w \times 1000}{m \times W}$$

or

$$m = \frac{K_b w \times 1000}{\Delta T_b W}$$

$$m = \frac{(3.9 \text{ K kg mol}^{-1})(5.141 \times 10^{-4} \text{ kg})(1000 \text{ g kg}^{-1})}{(0.325 \text{ K})(35 \times 10^{-3} \text{ kg})}$$

$$= 176.3 \text{ g mol}^{-1} = 0.1763 \text{ kg mol}^{-1}$$

Problem 4:

18.2 g of urea is dissolved in 100 g of water at 50°C. The lowering of vapour pressure produced is 5 mm Hg. Calculate the molecular weight of urea. The vapour pressure of water at 50°C is 92 mm Hg.

Solution:

Since the solution is to very dilute, the complete Raoult's Law Equation applied is

$$\frac{p - p_s}{p} = \frac{w / m}{W / M + w / m} \quad ...(1)$$

In this case:

w, the weight of solute (urea) = 18.2 g

W, the weight of solvent (water) = 100 g

m, the mol wt of solute (urea) = ?

M, the mol wt of solvent (water) = 18

$p - p_s$, the lowering of vapour pressure = 5 mm

p, the vapour pressure of solvent (water) = 92 mm

Substituting these values in equation (1)

$$\frac{5}{92} = \frac{18.2/m}{18.2/m + 100/18}$$

Whence m, the molecular weight of urea = 57.05

Problem 5(a):

A current of dry air was passed through a solution of 2.64 g of benzoic acid in 30.0 g of ether ($C_2H_5OIC_2H_5$) ad the through pure ether. The loss in weight of the solution was 0.645 g and of the ether 0.0345 g. What is the molecular weight of benzoic acid?

Solution:

According to the theory of Ostwald-Walker method,

$$\frac{p - p_s}{p} = \frac{W_2}{W_1 + W_2} \qquad ...(1)$$

In this case,

w_1, loss of weight of solution = 0.645 g

w_2, loss of weight of solvent = 0.0345 g

Substituting values in equation (1)

$$\frac{p - p_s}{p} = \frac{0.0345}{0.645 + 0.0345} = \frac{0.0345}{0.6795}$$

From Raoult's Law, we have

$$\frac{p - p_s}{p} = \frac{w/m}{w/m + W/M} \qquad ...(2)$$

M, the molecular weight of ether, $(C_2H_5)_2O$ = 48 + 10 + 16 = 74

Substituting values in (2)

$$\frac{0.0345}{0.6795} = \frac{2.64/m}{2.64/m + 30/74}$$

Whence m, the molecular wt of benzoic acid = 122.

Problem 5(b):

Air was passed through a solution containing 2 × 10 ² kg of a substance in 0.1 kg of water and the through pure water. The loss in mass of the solution was 2.945 × 10 ³ kg ad that of pure water was 5.6 × 10 ⁵ kg. Calculate the molar mass of the substance.

Solution:

Here $(p - p_s) \propto$ Loss in mass of pure water

$p_s \propto$ Loss in mass of solution

Therefore $p - p_s \propto 5.9 \times 10^{-5}$ kg.

$p_s \propto 2.945 \times 10^{-3}$ kg

$p \propto (5.9 \times 10^{-5} + 2.945 \times 10^{-3})$ kg

$\propto 3.004 \times 10^{-3}$ kg

$w = 2 \times 10^{-2}$ kg

$m = 18 \times 10^{-3}$ kg mol^{-1}

$W = 0.1$ kg

$M = ?$

Since $$\frac{p - p_s}{p} = \frac{wM}{Wm}$$

or $$m = \frac{pwM}{(p - p_s)W}$$

$$= \frac{(3.004 \times 10^{-3}\,\text{kg})(2 \times 10^{-2}\,\text{kg})\,(18 \times 10^{-3}\,\text{kg mol}^{-1})}{(5.9 \times 10^{-5}\,\text{kg})(0.1\,\text{kg})}$$

$= 0.183$ kg mol^{-1}

Problem 5(c):

In Ostwald-Walker experiment, air was blown through a solution containing a certain amount of solute (M = 0.2785 kg mol ¹) in 15 × 10 ² kg of water and then through pure water. The loss in mass of water was found to be 8.27 × 10 ⁵ kg while the mass of water absorbed in sulphuric acid tube was 3.317 × 10 ³ kg. Calculate the amount of solute.

Solution:

$p \propto$ Mass of water absorbed by sulphuric acid

$$p - p_s \propto \text{Loss in mass of water}$$

$$p \propto 3.317 \times 10^{-3}\ \text{kg}$$

$$p - p_s \propto 8.27 \times 10^{-5}\ \text{kg}$$

$$m = 0.2785\ \text{kg mol}^{-1}$$

$$w = ?$$

$$M = 18 \times 10^{-2}\ \text{kg}$$

$$W = 15 \times 10^{-2}\ \text{kg}$$

We know

$$\frac{p - p_s}{p} = \frac{wM}{mW}$$

or $$w = \frac{(p - p_s)\,mW}{pM}$$

$$w = \frac{(8.27 \times 10^{-5}\ \text{kg})(0.2785\ \text{kg mol}^{-1})(15 \times 10^{-2}\ \text{kg})}{(3.317 \times 10^{-3}\ \text{kg})(18 \times 10^{-3}\ \text{kg mol}^{-1})}$$

$$= 5.7 \times 13^{-2}\ \text{kg}$$

Problem 5(d):

A sample of camphor used in the Rast method of determining molecular weights had a melting point of 176.5°C. The melting point of solution containing 0.522 g camphor and 0.0386 g of an unknown substance was 158.8°C. Find the molecular weight of the substance. K_f of camphor per k is 37.7.

Solution:

Applying the expression

$$m = \frac{1000 \times K_f \times w}{\Delta T \times W}$$

In present case, we have

$$\Delta T = 176.5 - 158.5 = 17.7$$

$$K_f = 37.7$$

$$w = 0.0386$$

$$W = 0.522$$

Substituting these values

$$m = \frac{1000 \times 37.7 \times 0.0386}{17.7 \times 0.522} = 157.$$

Problem 6:

The lowering of freezing point of benzene was 2.33 K when 4.12 × 10^{-4} kg of a solute of unknown molar mass was dissolved in 9.13 × 10^{-3} k of benzene. Calculate the molar mass of the solute. Molal depression constant for benzene is 5.1 K kg mol^{-1}.

Solution:

Here,

$$\Delta T_f = 2.33 \text{ K}, \qquad K_f = 5.1 \text{ K kg mol}^{-1}$$

$$m = ? \qquad W = 9.31 \times 10^{-3} \text{ kg}$$

$$w = 4.12 \times 10^{-4} \text{ kg}$$

We know,

$$\Delta T = K_f \frac{w \times 1000}{m \times W}$$

or

$$m = \frac{K_f \times w \times 1000}{W \times \Delta T_e}$$

$$= \frac{(5.1 \text{ K kg mol}^{-1})(4.12 \times 10^{-3} \text{ kg})(1000 \text{ g kg}^{-1})}{(9.31 \times 10^{-3} \text{ kg})(2.33 \text{ K})}$$

$$= 96.87 \text{ g mol}^{-1}$$

$$m = 9.687 \times 10^{-2} \text{ kg mol}^{-2}$$

Problem 7(a):

A brass sample composed of 20% zinc and 80% copper by mass metls at 1268 K. Pure copper melts at 1357 K. What is the molal freezing point constant for copper? (Atomic mass of zinc is 65g mol^{-1}).

Solution:

$$\Delta T_f = (1357 - 1268) \text{ K} = 89 \text{ K}, \; W = 80 \times 10^{-3} \text{ kg}$$

$$m = 65 \times 10\text{– }3 \text{ kg mol}^{-1}, \quad w = 20 \times 10^{-3} \text{ kg}$$

$$= 65 \text{ g mol}^{-1}$$

Since $\Delta T_f = K_f \dfrac{w \times 1000}{m \times W}$

or $$K_f = \frac{\Delta T_f \times W \times m}{w \times 100}$$

$$= \frac{(89\text{ K})(80 \times 10^{-3}\text{ kg})(65\text{ g mol}^{-1})}{(20 \times 10^{-3}\text{ kg})(1000\text{ g kg}^{-1})}$$

$$= 23.14\text{ K kg mol}^{-1}$$

Problem 7(b):

0.440 g of a substance dissolved in 22.2 g of benzene lowered the freezing point of benzene by 0.567°. Calculate the molecular weight of the substance. (K_f = 5.12°C mole^{-1}).

Solution:

We can find the molecular weight by applying the expression

$$m = \frac{100 \times K_f \times w}{\Delta T_f \times W}$$

In this case

$$w = 0.440$$

$$\Delta T = 0.567$$

$$W = 22.2$$

$$K_f = 5.12$$

Substituting the values,

$$m = \frac{1000 \times 5.12 \times 0.440}{0.567 \times 22.2} = 178.9$$

∴ Molecular weight of substance = 178.9.

Problem 7(c):

A solution of 0.124 g of a substance, X, in 25.0 l of ethanoic acid (acetic acid) has a freezing point 0.324°C below that of the pure acid 16.6°C. Calculate the molecular weight (relative molecular mass) of X, given that the specific latent heat of fusion of ethanoic acid is 180.75 J g^{-1}.

Solution:

Calculation of Molal depression Constant

We know that

$$K_f = \frac{RT_f}{L_f} \qquad ...(1)$$

Here, freezing point of benzene,

$$T_f = 273.2 + 16.6 = 289.9 \text{ K}$$

Specific latent heat of fusion,

$$L_f = 180.75 \text{ J g}^{-1}$$

Substituting in the equation (1)

$$K_f = \frac{8.314 \times (289.9)^2}{180.75} = 3.86°$$

Problem 8:

A solution of 1.0 × 10 ²kg of sodium chloride in 1000 g of water freezes at – 0.604°C. The molal depression constant K_f of water is (1.85 K kg mol ¹). Calculate the degree of dissociation of sodium chloride.

Solution:

$$(\Delta T_f)_{cal} = K_f \frac{w \times 1000}{W_m}$$

$$= \frac{(1.85 \text{ deg kg mol}^{-1})(1 \times 10^{-2} \text{ kg})(1000 \text{ g kg}^{-1})}{(1000 \text{ g kg})(58.5 \times 10^{-3} \text{ kg mol}^{-1})}$$

$$= 0.316°C$$

$$(\Delta T_f)_{obs} = 0.0 - (-0.604) = 0.604°C$$

van't hoff factor,

$$i = \frac{(\Delta T_f)_{obs}}{(\Delta T_f)_{cal}} = \frac{0.604}{0.316} = 1.91$$

Sodium chloride dissociates as	$NaCl$ ⇌	Na^+ +	Cl^-
Number of moles initially	1	0	0
Number of moles after dissociation	1 – α	α	α

Total number of moles after dissociation = 1 – α + α + α = 1 + α

$$i = \frac{\text{Number of moles after dissociation}}{\text{Number of moles initially present}} = \frac{1 + \alpha}{1}$$

Therefore,

$$\frac{1+\alpha}{1} = 1.91$$

or $$\alpha = 0.91 \quad \text{to } 91\%$$

The degree of dissociation = 91%.

Problem 9:

A solution containing 0.3×10^{-3} kg of benzoic acid ($M = 122 \times 10^{-3}$ kg mol^{-1}) in 2×10^{-2} kg of benzene freezes at 0.317°C below the freezing point of the solvent. Calculate :

(i) the degree of association assuming that the acid exists as dimmer in benzene, and

(ii) the apparent molar mass of the acid; K_f for benzene is 5.1.

Solution:

$$(\Delta T_f)_{cal} = K_f \frac{w \times 1000}{mW}$$

$$= \frac{(5.1 \text{ deg kg mol}^{-1})(0.3 \times 10^{-3} \text{ kg})(100 \text{ g kg}^{-1})}{(122 \text{ g mol}^{-1})(2 \times 10^{-2} \text{ kg})}$$

$$= 0.627°\text{C}$$

The observed value of ΔT_f = 0.317°C.

Hence van't Hoff factor i is given as

$$i = \frac{(\Delta T_f)_{obs}}{(\Delta T_f)_{cal}} = 0.506$$

From equation (3), α is given by

$$\alpha = \frac{i-1}{\frac{1}{n}-1} = \frac{0.506-1}{\frac{1}{2}-1} = 98.8\%$$

(ii) Since $$i = \frac{(M)_{cal}}{(M)_{obs}}$$

$$0.506 = \frac{122}{(M)_{obs}}$$

$$(M)_{obs} = \frac{122}{0.506} = 241 \text{ g mol}^{-1}.$$

3

VISCOSITY AND CHEMICAL CONSTITUTION

INTRODUCTION

Viscosity has been found to be neither additive nor a constitutive property and therefore it could not be used in deciding between different possible structures of molecules to the same extent as molar volume or parachor could be used. However, it was Dunstan (1909) who discovered an interesting relationship between viscosity and molar volume of non-associated liquids as follows:

$$\frac{d}{M} \times \eta \times 10^6 = 40 \text{ to } 70 \quad \ldots(1)$$

Table 1 : Values of $\frac{d}{M} \times \eta \times 10^6$.

Liquid	$(d/M) \times \eta \times 10^6$	Inference
Acetone	43	Unassociated
Toluene	56	Unassociated
Benzene	73	Unassociated
Water	559	Associated
Glycol	2750	More associated
Glycerol	116400	Highly associated

where d is the density, M the molecular weight and η the coefficient of viscosity of the liquid. This relationship has been employed to know whether a liquid is associated or not. If the value exceeds 70, the liquid exists as associated molecules. Table 1 shows how the reference can be drawn about the nature of liquids. The values of acetone, toluene and benzene have been 43 56 and 73 respectively, thereby indicating that these liquids are undissociated. On the other hand, the values for water, glycol and glycerol have been 559, 2750 ad 116400 respectively thereby indicating that these liquids are associated and further indicating that glycerol has been much more associated than glycol.

Molecular Viscosity or Molar Viscosity

As M/d represents molecular volume, it means that $(M/d)^{2/3}$ represents the molar surface area. The product of molar surface and viscosity is termed as molar viscosity.

Molar viscosity = molar surface 0215 viscosity

$$= \left(\frac{M}{d}\right)^{2/3} \times \eta$$

It was found by Thorp and Rodge (1894) that molecule viscosity is an additive and constitutive property at the boiling point of liquid. Values for different atoms and linkages have been determined and given in standard charts. From these the molar viscosity for a compound can be calculated. If the calculated and experimental values come out to be the same, this is taken as a confirmation of the structure of the substance. The values of molar viscosity contributions by different atoms have been given as follows:

Atom	H	C	O (in OH)	O (in — CO)	S
Molar viscosity	80	– 98	196	248	155

Rheochor

According to Newton Friend, the product of molar volume and the eighth root of viscosity is constant at least at the boiling point *i.e.,*

$$\frac{M}{d} \times \eta^{1/8} = R$$

where 'R' is a constant called the *Rheochor*. Rheochor, like molar viscosity, is also an additive and constitutive property. Physical significance of rheochor is that it may be regarded as the molar volume of a liquid at a temperature when its velocity is unity. *Rheochor is not very useful for determining the chemical constitution of a substance.*

The atomic and structural rheochor values (like parachor values) are given in Table 2.

Table 2 : Atomic and Structural Rheochors

Atom	Rheochor	Linkage	Rheochor
Carbon	12.8	Covalent bod	0.0
Oxygen (in ether)	10.0	Coordinate bond	0.4
(in ketone)	13.2		
Hydrogen (in C—H)	5.5	6-membered ring (hat)	– 5.6
(in C—OH)	10.0	$—CH_2$	23.6
(in HCl)	9.7	$—C_6H_5$	100.7
(in HBr)	12.6	$—C(=O)O$	36.0
(in HI)	15.0	$—NH_2$	20.6
Chlorine	27.3	>NH	13.6
Bromine	35.8	CN	33.0
Iodine	47.6		
Nitrogen	6.6		

It is to be made clear that the rheochor is less useful for determining chemical structures that parachor ad is not used frequently for studying the chemical constitution.

REFRACTIVE INDEX

If light is allowed to pass from a rarer medium (*say air*) into a denser medium (*say a liquid*), it is bent or refracted towards the normal. The ratio of the sine of the angle of incidence and that of refraction is constant and characteristic of that liquid (Snell's law). The constant ratio is termed as the refractive index of the liquid ad may be put as follows:

$$\mu = \frac{\sin i}{\sin r}$$

According to the wave theory of light, the ratio of the sines of the angles of incidence and refraction is identical with the ratio of the velocities of light in the two media. Hence, refractive index may be defined as follows:

Thus,
$$\mu = \frac{\sin i}{\sin r} = \frac{\text{Velocity of light in air}}{\text{Velocity of light in liquid}}$$

If a ray of light is allowed to pass from a rarer to denser medium, it can be shown from the law of refraction that

$$\frac{\sin i}{\sin r} = \frac{\mu_2}{\mu_1}$$

Here μ_1 is the index of refraction of the rarer and μ_2 the index of refraction of the denser medium. The angle of incidence can never be evidently greater than a 90° and if it is 90° the above equation gets reduced to $\sin r = \frac{\mu_1}{\mu_2}$ or sin 90° = 1.

MEASUREMENT OF REFRACTION INDEX

For the rapid determination of the refractive index of liquids, a number of instruments called refractometers have been constructed. We give below a brief account of the most prominent them which are :

(i) the Pulfrich refractometer, and

(ii) the Abbe refractometer.

1. Pulfrich Refractometer

Pulfrich refractometer is very accurate and simple in principle and is depicted. The essential part of the instrument is a right angled glass prisms with a small glass cell cemented to its top. The liquid to be examined is kept in the cell and a beam of monochromatic light is made to enter the liquid at 'grazin incidence' along the surface between the liquid and the prism. It follows the path of *ABCD* and is observed in a telescope at *D*. If the telescope is moved to make an angle with the horizontal which is less than i, no light could reach it. A very accurate determination can, therefore, be made of the angle i at which a sharp boundary between a dark and light field can be seen through the telescope. For a ray of light passing from the liquid into the prism; if r refers to the angle of refraction when the angle of incidence is 90°, we have already stated that

$$\sin r = \frac{\mu_1}{\mu_2} \qquad ...(1)$$

where μ_1 is the refractive index of the liquid and μ_2 is that of the glass prism. It is also clear from the diagram that

$$\frac{\sin i}{\sin(90° - r)} = \mu_2 \quad ...(2)$$

Bus sin (90° – *r*) = cos *r*, we have

$$\frac{\sin i}{\cos r} = \mu_2$$

or

$$\cos r = \frac{\sin i}{\mu_2} \quad ...(3)$$

But

$$\sin r = \sqrt{1 - \cos^2 r} \quad ...(4)$$

On substituting the value of cos r from (4), we obtain

$$\sin r = \sqrt{1 - \frac{\sin^2 i}{\mu_2^2}}$$

and from Eq. (1), $\mu_1 = \mu_2 \sin r$

Hence

$$\mu_1 = \sqrt{\mu_2^2 - \sin^2 i}$$

If the refractive index μ_2 of the glass is known and angle i is measured, μ_1, the refractive index of the liquid, can be calculated. In actual practice, however, it is not necessary to go through the above calculations because the makers of the instrument supply tables giving refractive indices (μ_1) corresponding to different values of i.

The index of refraction depends upon the wave-length of light employed, the index for red rays being less than that for the violet rays. Measurements of the refractive index referred to D-line of sodium are usually indicated by the symbol μ.

2. Abbe's Refractometer

The principle of this instrument happens to be the same as that used in the Pulfrich refractometer. This instrument is, however less accurate than the Pulfrich refractometer. It is principally designed for the rapid determinations of the refractive index of small quantities of liquids.

(a) *Construction* : A general idea of the construction of this instrument.

A and B are two glass prisms. The hypotenuse surface of b is polished while that of A is finely ground. The two prisms are housed in metal casings hinged at H. The two prisms faces can be held i contact with clamp C. A and B can be rotated about a horizontal axis immediately beneath a telescope T. To the metal case carrying the prism, is attached arc arm, R which is able to move along a graduated scale S, the reading on which provides directly the refractive index.

Working

A drop of the liquid is kept upon the surface of prism A. On clamping the two prisms A and B a film of the liquid spreads between them. Light reflected by mirror M is then made to direct towards the prism system. On reaching the ground surface of A it gets scattered into the liquid film. No ray can, however, enter B with a greater angle of refraction than that of the ray corresponding to '*grazing incidence*'. Hence when viewed through a telescope, just as in the case of Pulfrich refractometer the field of view gets divided into bright and dark portions. The edge of the bright portion when coincided with the cross-wire of the telescope provides the refractive index on the scale. If white light is employed, as is the case in practice, a diffused coloured border is seen in the telescope. This is made sharp and the colours are removed by the adjustment of two prisms, (not shown in the diagram) attached at the nose of the telescope.

As temperature control is of great importance in determining the refractive index of liquids, the prisms A and B are enclosed in a water jacket, J so as to maintain a constant temperature.

Plane Polarised Light

A beam of light, proceeding towards this paper and at right angles to the paper, possesses wave components that are vibrating in all possible planes passing through the axis along which the beam is travelling (*axis of propagation*). If a beam of light is passed through a properly oriented crystal of the mineral Iceland spar; or calcite (calcium carbonate) the beam is split into two beams and the process is said to be double refraction. The two emerging beams both have only a single plane of vibration and the plane of vibration in beam *a* is perpendicular to that of beam *b*. In short the crystal is able to separate each wave plane of vibration into horizontal and vertical components b bending one component more that the other. A beam of light having a single plane of wave vibration is said to be plane polarised as represented by either beam 'a' and 'b'.

The generation of a single beam of plane-polarized light may be accomplished b separation of the two beams a and b. A suitable apparatus for this operation was invented in 1828 by William Nicol, a Scotish

Physicist known as a '*Nicolprism*'. The apparatus consists of two crystals of Iceland spar cut to certain angles and cemented by canada balsam. A beam of light entering at the left is doubly refracted into beams a and b. Beam 'a' passes through both crystals, emerging plane polarized on the right. Because of the cut of the crystal ad the index of refraction of the canada balsam; beam b is reflected at the crystal juncture and does not enter the second crystal. Plane polarised light may also be generated by a polaroid, a cooperatively recent invention of an American E. H. Land.

Certain crystals said to have the property of *di-chorism*, will absorb light components vibrating in one plane more strongly that those vibrating in the perpendicular plane. Thus a suitable thickness of such a crystal can be used to produce plane-polarised light. Polaroid consists of certain dichroic derivatives of quine the crystals of which are properly oriented and embedded in a transparent plastic.

NATURE OF OPTICAL ACTIVITY

When a plane polarised light passes through certain organic substances it changes its direction of motion. This phenomena of rotation of plane of polarised light b certain substances is called "*optical activity*" and the substance which rotates the plane of polarised light is called "*optical active.*"

Optically active compounds can be divided into two categories: one in which the activity is evident in the solid state only, *i.e.,* due to the crystalline structure of the compounds, and the other in which optical activity is shown by the compounds in the solid, pure liquid, solution or gaseous forms. As the optical activity of the former class has been related to the disappears, *i.e.,* when the substance melts.

This type of optical activity is found with quartz, sodium chlorate etc. Optical activity of the latter class has been related to the molecular structure of the compounds, *i.e.,* this property depends on the arrangement of the atoms in a molecule. As the molecular structure does not get altered b fusion dissolution, or vaporization, therefore, the property gets manifested equally well in these forms.

DEXTRO AND LAEVO ROTATORY SUBSTANCES

The rotation of the plane of polarised light ay take place either towards the right or the left.

Rotation towards right : An optically active substance which rotates the plane of the polarised light towards the right is called dextro rotator (d).

Rotation towards left : An optically active substance which rotates the plane of the polarised light towards the left is called *laevo rotator* (l).

The angle of rotation depends upon the following factors:

(i) Nature of the substance

(ii) Thickness of the layer through which the light passes.

(iii) Wave length of the light used.

(iv) Density of the solution.

(v) Temperature of the experiment.

(vi) Nature of the solvent.

Specific Rotation

If 'R' is the observed rotation 'λ' the wave length of light used, 't' the temperature of the solution 'd' is density and 'T' the length of the column of the solution through which light passes, then

$$\frac{R}{l.d} = [\alpha]_{\lambda}^{t} \quad ...(1)$$

where $[\alpha]_{\lambda}^{t}$ is a constant called the *specific rotation* of the substance is solution for the light of wavelength λ at the temperature 't'.

For practical purposes, sodium vapour lamp is used as the source of light. Then λ is replaced by 'D' the wavelength of 'D' line of sodium. Equation (1) is then written as

$$[\alpha]_{D}^{t} = \frac{R}{Id} \quad ...(2)$$

when $d = 1$ gm/c.c.

and $I = 1$ decimetre

then $[\alpha]_{D}^{t} = R$ ...(3)

Hence the specific rotation of a substance may be defined *as the angle of rotation produced when the plane polarised light passes through one decimetre length of a solution containing 1 cm per c.c. of the optically active substance.* Now if the solution contains 'C' gms. of the substance dissolved in 100 c.c. of water then,

$$\alpha = \frac{C}{100} \qquad ...(4)$$

Substituting in equation (2), we get

$$[\alpha]_D^t = \frac{100 \cdot R}{l \cdot c} \qquad ...(5)$$

The specific rotation is a characteristic of substance.

Molecular Rotation

Optical rotatory power of a compound is sometimes given as a Molecular rotation [M]. This value is by definition the molecular weight m of the substance multiplied by the specific rotation and divided by 100 to reduce the size of the number.

$$[M] = \frac{[m]\alpha}{100}$$

Measurement of Specific Rotation—The Polarimeter

The instrument used for measurement of specific rotation is called a polarimeter. It consists of two nicol prisms A and P. The prism P is called the polariser and the prism A is called the analyser.

Light from source S is allowed to fall on the polariser and it gets plane polarised. If the axis of the analser is parallel to that of P only then light can be viewed through it. But if the axis of A is at right angles to that of P completely dark view is seen. To begin with the axis of A is adjusted at right angles to that of P by moving it on the graduated scale C so that a completely dark view is seen. The solution of known concentration of the optically active substance is placed in the tube in between the two prisms. It rotates the polarised light through a certain angle. To make the field of view dark again the analyser has to be rotated through the same angle. This gives the observed angle of rotation 'R' from which the specific rotation is calculated with the help of equation (5) given above. The magnitude of rotation of plane-polarised light by an optically active substance will depend on several factors.

Different wavelengths of light used in the polarimeter will produce different rotations. It is, therefore, necessary to use a known monochromatic (narrow wavelength range) light source. The yellow sodium D-line (5890-5896 Å) is commonly used.

Magnitude of rotation depends upon the number of molecules of optically active substance in the light path ad thus on the concentration of an solution used, and the length of the sample cell.

Other significant variables are the temperature and solvent used (if any) for the optically active substance.

Chemical Constitution and Optical Activity

It has been found the optical activity is a constitutive property, *i.e.*, it depends upon the arrangement of atoms with the molecule. The organic compounds having at least one asymmetric carbon atom are found to possess this property. *Asymmetric* carbon is that which has all its found velencies satisfied by four different groups or atoms. Compounds contain asymmetric nitrogen or silicon atoms are also found to exhibit this property. Further, it has been found that a substance with optical activity always exists in these forms. Two of them are rotating the plane of polarized light in opposite directions viz., left and right and the third being optically inactive. Thus if a substance shows optical activity it must have a asymmetric atom in it. For example lactic acid has the formula

```
H   CH3
 \ /
  C
 / \
OH  COOH
```

and is found to be optically active. Asymmetric carbon atom is designated by an asterisk.

According to Le-Bell and Van't Hoff, all the four groups attached to the carbon atom do not lie in the same plane as the carbon atom, but are situated at the four corners of a regular tetrahedron.

So accordingly the arrangements of various groups in space, the lactic acid molecule can be represented in two different ways do not have a plane of symmetry.

In other words it is not possible to pass a plane through any of the two tetrahedrons so as to divided them into exactly identical half. Such a carbon atom is called an asymmetric carbon atom.

OPTICAL ISOMERS AS MIRROR IMAGE OF EACH OTHER

These two models are not super-impossible, but are related to each other as an object to its mirror image. The exact mechanism by which the molecule rotates the plane of polarized light is not known.

It can, however that in one case the groups are arranged clockwise while in the other anti-clockwise. One of these represents the laevo farmhand the other dextro form.

Recently optical rotator dispersion curves have been used in the study of the stereochemistry of optically active ketones such as cyclohexanone derivatives.

The optical activity of a compound having more than one asymmetric C-atom depends on the arrangement of the groups around each C-atom. Suppose a compound is having two asymmetric C-atoms. If both of them rotate the plane of polarization towards the same direction, the optical activity is enhanced, whereas if one rotates the plane in an opposite direction to the other, then the activity is reduced.

The latter is an example of internal compensation. If the groups around the two C-atoms are identical, the internal compensation results in zero activity and such a form is termed as the *meso-form.*

Thus in tartaric acid ($CH_2HCHOHCHOHCO_2H$) we have the *d, l*, the meso and the racemic forms. Each asymmetric C-atom represents one active centre. This active centre gives rise to two possibilities of optical rotation either towards right or towards left.

Hence a compound having *n*-asymmetric C-atoms will give rise to 2^n such possibilities and hence will have 2^n isomers, *e.g.*, glucose, $C_6H_{12}O_6$, has four asymmetric C-atoms ad hence it must give rise to 16 isomeric hexose sugars.

PRINCIPLE OF OPTICAL SUPERPOSITION

J.H. Van't Hoff gave the principle of optical superposition which is applicable possible to isomeric substances having several asymmetric carbon atoms.

It may be stated as "*the rotating power of substances possessing several asymmetric carbon atoms refers to the algebraic sum of the contributions of each separate carbon atom which is a definite amount and independent of the configurations of the other atom.*" This can be understood taking the examples of four pentose sugars.

```
        |                          |
  HO—C—H + A              HO—C—H + A
        |                          |
  HO—C—H + B              HO—C—H + B
        |                          |
  HO—C—H + C               H—C—OH—C
        |                          |
        I                          II
```

```
      |                            
HO—C—H + A              H—C—OH—A
   |                       |
 H—C—OH—B              HO—C—H + B
   |                       |
 H—C—OH—C               H—C—OH—C
   |                       |
  III                      IV
```

According to the principle the specific rotation of II should be the sum of the values for the other three isomeric sugars:

Hence + A + B – C = (+ A + B + C) + (+ A – B – C)

+ (– A + B – C)

Optical activity is used to confirm the structure but it is not able to disapprove any structure.

Optical Rotatory dispersion

In 1817 Biot reported that the magnitude of optical rotation gets changed with the change of wave length of the light used in the determination. The variation of the optical activity with the frequency of the light is termed as *optical rotatory dispersion.* It can be interpreted in terms of variation of the polarizability of the molecules in light of different frequencies. P. Drude deduced the following equation which gives the wavelength dependence of a great umber of optically active substances.

$$\alpha = \frac{K}{\lambda^2 - \lambda_0^2}$$

where λ_0 and *K* are characteristics of the substance.

If [α] is plotted against wave length, the following two types of the curves are obtained..

(i) A curve showing a smooth increase in rotation (+ or –) with decreasing wave length,

(ii) A curve having peak and trough.

The curve (i) exhibits peak at a longer wavelength. This is termed as *positive cotton effect* curve. The curve B in which the trough is found at longer wavelength is termed as *negative cotton effect curve.* Optical rotatory dispersion studies are usually carried out in UV visible range when a functional group of the compound has a weak absorption band. It has been successfully used in ascertain in the structures of complicated optically active compounds.

Molecular Refraction

The refractive index of a fluid varies with temperature and pressure, as these factors alter the number of molecules i, the path of the light. Hence, it cannot be used to compare refractive powers of different fluids in relation to their molecular structures.

To eliminate the effect of these factors, H.A. Lorenz (1880) from the electromagnetic theory of light, and L.V. Lorenz (1880) from the wave theory of light, independently deduced the following theoretical relationship between refractive index and density, which should be constant at all temperatures:

$$R = \frac{\mu^2 - 1}{\mu^2 - 2} \times \frac{1}{d} \qquad ...(1)$$

where μ is the refractive index of a liquid and 'd' its density, R is a constant called the *specific refraction or refractivity*. Multiplying the above equation by , the molecular weight of the liquid on both sides we get,

$$R \times M = \frac{\mu^2 - 1}{\mu^2 - 2} \cdot \frac{M}{d} = R_M \qquad ...(2)$$

The product R × M is denoted by R_M and is called the *molar refraction or molar refractivity*. R_M is characteristic of the liquid and remains constant at the given temperature.

The molar refraction is independent of temperature and pressure, but as is the case with refractive index it varies with the wave-length of the light used.

Generally, the D-line of sodium is the light source. If any other source is employed (*e.g.*, the D-line in the hydrogen spectrum) it should be specifically mentioned.

The molar refractivity of a solute, solvent and solution can be determined as follows:

A know weight of the solid is dissolved in a known weight of a suitable solvent. The refractive index and density of the solution are determined in the usual way. If R' is molar refractivity of the solution, it is given by,

$$R' = \frac{\mu^2 - 1}{\mu^2 - 2}\left(\frac{N_1M_1 + N_2M_2}{d}\right) \qquad ...(3)$$

where μ is the refractive index of the solution N_1, N_2 the mole fractions of the solute and solvent respectively, M_1 and M_2 are their molecular weights and 'd' and density of the solution. The molar refractivity of the solution R′, is related to the molar refractivities of the solute and solvent as:

$$R' = N_1R_1 = N_2R_2 \qquad ...(4)$$

where R_1 ad R_2 are the molar refractive is of the solute and solvent respectively ad N_1 and N_2 have the same meaning as before.

The refractive index and density of the solvent are now determined and its molar refractivity R_2 is calculated with the help of equation (2). Substituting this value is equation (4). R_1 the molar refractive of the solutes is calculated.

REFRACTIVITY AS AN ADDITIVE AND CONSTITUTIVE PROPERTY

It was found by comparing the R_M values for different substances that it is both additive and constitutive property, *i.e.*, it depends upon both the number and arrangement of atoms in the molecule. Thus from a study of molar refractions of a large umber of compounds the constants for different atoms and types of linkages were determined. For example, in a homologous series of aliphatic compounds of difference of CH_2 in composition makes a difference of 4.618 in molecular refractivity. The contribution of other atoms and structures can be calculated like molar volume as follows:

$$\text{RM for } C_7H_{16} = 34.54 \text{ cm}^3 \text{ mol}^{-1}$$

$$\text{RM for } C_6H_{14} = 29.92 \text{ cm}^3 \text{ mol}^{-1}$$

$$\text{Difference}(= CH_2) = 4.62 \text{ cm}^3 \text{ mol}^{-1}$$

$$\text{Thus,} \quad C_6H_{14} = 6CH_2 + 2H$$

$$29.92 = 6 \times 4.62 + 2H$$

$$\therefore R_M \text{ for H atom} = 1.1 \text{ cm}^3/\text{ atom}$$

$$\text{C atom} = 4.62 - 2 \times 1.10$$

$$= 2.42 \text{ cm}^3/\text{g atom}$$

The molar refraction values for some elements and groups based of the refractive indices measured with the D-line of sodium have been included in Table 3.

Table 3 : Molar refraction contributions of atoms and bonds (Vogel 1948)

Atom or structure	$R_M (cm^3)$	*Atom or Structure*	$R_M (cm^3)$
H	1.028	C—O	4.601
C	2.591	—OH	1.546
Cl	5.844	—COOH	7.226
Br	8.741	$—NO_2$	6.713
I	13.954	=	1.575
O(>C=O)	2.573	≡	1.977
		6 membered ring	– 0.15
O(R—O—R)	1.764	5 membered ring	– 9.89
O(— OH)	1.518	4 membered ring	0.317
C_3 radical	5.653		
C_2H_5 radical	10.300		

With the help of molar refraction contributions, given in Table 1.6 the molar refractivity of a compound o the basis of its assumed structure is calculated and compared with the value obtained experimentally. This can be understood from the following examples.

1. *Structure of benzene :* Kekule's formula for benzene is:

```
        CH
       /  \\
     HC    CH
     ||    |
     HC    CH
       \  //   c
        C
        |
        H
```

There are six carbon atoms, 6 hydrogen atoms, 3 double bonds and one six carbon atom ring.

∴ Molar refraction due to 6 C atoms $= 6 \times 2.591 = 15.546$

Molar refraction due to 6 H atoms $= 6 \times 1.028 = 6.168$

Molar refraction due to 3 double bonds $= 3 \times 1.575 = 4.725$

Molar refraction due to one six membered ring $= 1 \times (-0.15) = -0.15$

Total 25.289

The experimental value 25.95 is fairly close to the above calculated value. Hence the Kekule's structure for benzene is correct.

2. *Structure of acetylene dibromide* : The structure of acetylene dibromide is

$$BrCH = CHBr$$

	2C = 2 × 2.591	= 5.182
	2H = 2 × 1.028	= 2.056
	2Br = 2 × 8.741	= 17.482
One double bond	= 1 × 1.575	= 1.575
	Total	= 26.295

The observed value is 26.30. Hence the formula of acetylene dibromide is verified. Thus molecular refractivities have been useful in deciding between the alternate structures of the various isomeric compounds.

3. *Structure of acetic acid* : The correct structure of acetic acid can be either (I) or (II).

	$CH_3.\overset{\overset{O}{\|\|}}{C}.OH$ (I)	$CH_3.COOH$ (II)
2C	3 × 2.591 = 5.182	CH3 radical = 5.653
4H	4 × 1.028 = 4.112	— COOH = 7.226
O(C=O)	1 × 2.573 = 2.573	Total = 12.879
— OH	1 × 1.518 = 1.518	
	Total 13.385	

Observed value of R_M = 13.021.

As the calculated value of R_M of structure I has been found to be very close to observed value of R_M, the structure I is the correct structure of acetic acid.

4. *Keto-enol tautomerism* : Molar refraction can give an approximate indication whether one or the other form in keto-enol tautomerism of a compound predominates in its normal state.

The enol form is having double bond and therefore two forms should possess different values of R_M. The calculated molar refractions of the keto and enol forms of acetoacetic ester are 31.57 and 32.62, respectively, hence

$$\underset{\text{Keto}}{\underset{31.57}{CH_3.CO.CH_2.COOC_2H_5}} \qquad \underset{\text{enol}}{\underset{32.62}{CH_3.\overset{\overset{OH}{|}}{C}{=}CH.COOC_2.I_5}}$$

The enolic form is having conjugated double bond and therefore, an optical exaltation equal to 1.8 is found. The molar refraction of the enolic form should be 34.42. The measured molar refraction of the acetoacetic ester in its normal state is 32. The proportion of enolic form should be given as follows:

$$\frac{32.00 - 31.57}{34.42 - 31.57} \times 100 = 15\%$$

Molecular refraction values such as parachor can be used for elucidation the chemical constitution. Even though the calculated and observed values of molecular refraction closely tally in many cases but anomalous results are also obtained. This is ascribed to the optical exaltation.

Optical Exhalation

It is observed that when a compound contains ore than on double bond, the molecular refractivity depends not only on the number of double bonds but on their) position also. Thus in case of conjugated system of double bonds (i.e., alternate double bonds there is a marked increase in the value of the observed molecular refractivity than the value calculated from the atomic and the structural constants. This behaviour is known as optical exhaltation.

For example, the observed R_M of isodially,

$$CH_3{-}CH{=}CH{-}CH{=}CH{-}CH_3$$

having conjugated double bonds, is 1.76 units higher than the calculated R_M whereas for the isomeric dually, $CH_2{=}CH{-}CH_2{-}CH_2{-}CH{=}CH_2$, in which the double linkages are not conjugated, the observed value of R_M is smaller by about 0.12 unit.

It has also been found that a carbonyl group, if present along with a double or triple bond, causes optical exaltation. Hence, the experimental value of molar refraction of phorone,

$$(CH_3)_2C{=}CH{-}CO{-}CH{=}C(CH_3)_2,$$

is 45.39 while the calculated value is 42.85.

$$\text{Optical exaltation} = 45.39 - 42.85 = 2.54$$

If the conjugated double linkages constitute a closed ring as in benzene and cyclo-octatetrene, the optical exaltation is to observed but if the conjugated system is partly within the ring and partly in a side chain as in styree I and acetophenone (II) the positive an only appears *i.e.,* optical exaltation is observed.

$C_6H_5-CH{=}CH_2$ (I) $C_6H_5-C(=O).CH_3$ (II) $-CH_3-C_6H_4-C_3H_7$ (III)

If there exists a gap in the ring as in α-phellandrene (III), the optical exaltation is present.

Refrachor

It is a new physical constant which relates refractive index μ_D^{20} with parachor. It is denoted by the symbol [F] and is expressed mathematically as follows:

$$F = -[P] \log (\mu_D^{20} - 1)$$

Refrachor can be used for elucidating structure but it has not found much application.

PHYSICAL PROPERTIES

Physical properties of a substance are defined as those properties *which could be studied and determined without causing any chemical change in it.* The physical properties like refractive index surface tension viscosity and dipole moment can be determined quantitatively with sufficient precision to serve as a tool of deductive information. Now, of course, we have far more suitable methods (*e.g.,* spectroscopic methods) available for this purpose.

The physical properties may be classified into various categories given below:

(i) Additive Property

It is the property which depends upon the nature and number of atoms present in a molecule of the compound, such that its magnitude is equal to the sum of the corresponding properties of the constituent atoms. For example, the molecular mass of a substance is equal to the sum of the atomic masses of the constituent atoms.

(ii) Constitutive Property

It is the property which depends entirely upon the ode of arrangement of atoms in the molecule, but independent of their number. The optical activity is one such property.

(iii) Additive and Constitutive Property

It is an additive property which also depends upon the manner in which the atoms are bonded to each other. For example, parachor is an additive as well as constitutive property.

A brief account of more important physical properties alongwith their applications to the problem of elucidating chemical structure has been dealt in this chapter.

Molar Volume of Liquids

The molar volume (V_m) of a substance may be defined as the volume (generally expressed in cm^3) occupied by one mole of the substance, under specified conditions of temperature and pressure. Mathematically, this can be put as follows:

$$V_m = \frac{\text{Molar mass}}{\text{Density}}$$

The units of molar volume are, those of volume per mole of the substance. It is possible to calculate the molar volume of a liquid from the value of the molar mass and density of the substance. The density can be measured by using a precalibrated density-bottle, or pyknometer. The density-bottle being convenient at room temperature, while pyknometer can be used for measuring density at various temperatures.

KOPP'S LAW

It is a well known fact that molar volume is equal to 22.4 litres in the case of all gases at N.T.P. On the basis of this numerous attempts were made to find out if the simple uniformity existed in liquids as well. One of the well known attempt was made by Kopp who measured densities of a large number of liquids at their boiling points which are approximately corresponding temperatures. Although he could not be able to obtain uniformity of results, yet he established a law called Kopp's law which may be stated as follows:

"*The molar volume of a substance is approximately equal to the sum of the atomic masses of its constituent atoms.*"

Kopp found the following regularities in the molar volumes of liquids:

(*i*) *The isomers belonging to the same homologous series are having nearly equal molar volumes*. An example is that the molar volume of *n*-heptane was 162.8 ml whereas that of *iso* heptane was 162.0 ml. Another example is that the of a volumes of normal and *iso*-butric acids were 108.2 ml and 108.9 ml respectively.

(*ii*) *Any two successive members of the same homologous series of organic compounds were found to differ in their molar volumes by about the same amount*. For example in the case of aliphatic alcohols, the difference in molar volumes of two successive members (such as methyl alcohol and ethyl alcohol) was 21.4 ml. Further, in the case of aliphatic acids, the difference in the values of two successive members was 22.5 ml. If similar other series such as paraffins aldehydes, amines, esters, etc., were considered, an average difference of 22.2 ml was found. These observations were found to form the basis of Kopp's law stated above.

Based on the studies of various compounds Kopps compiled a set of atomic volumes. Let Bas (1906) later on demonstrated that the molar volumes exhibited their dependence on the constitutional and structural influences. The following table records the atomic volumes given by Kopp as well as by Le Bas (1912).

Table 4 : Volume equivalents of elements

Atom	*Kopp*	*Le Bas*
H	5.5	3.7
C	11.0	14.8
Cl	22.8	22.2
Br	27.8	27.0
S	22.6	—
—O—	7.8	7.4
═O	12.2	12.0
I	37.5	37.0

From such data, it becomes possible to ascertain the molar volume of any organic substance and the structure of liquid molecule can be elucidated by comparing the calculated values from practically observed values of molecular volumes. To demonstrate this let us take the example of ethyl benzoate, $C_2H_5COOC_6H_5$.

LE BAS DATA

$$9C = 9 \times 14.8 = 133.2$$

$$10H = 10 \times 3.7 = 37.0$$

$$—O— = 1 \times 7.4 = 7.4$$

$$O = 1 \times 12.0 = 12.0$$

$$\text{(benzene ring)} = 1\ 0215\ (-15) = -15.0 \quad \text{(no included in table)}$$

$$174.6\ cm^3$$

Observed molecular volume = $174.6\ cm^3$.

Hence the structure of ethyl benzoate is $C_2H_5\overset{\overset{O}{\|}}{C}.\ OC_6.H_5$.

Limitations

However, it was soon found by various workers that molar volume was not pure an additive property. There occurred differences due to constitution as well. For example, atomic volume of oxygen was 12.0 in a carbonyl group ($>C{=}O$) but 7.8 in a hydroxyl group (— OH). It was also reported that the presence of benzene nucleus reduced the molar volume by 1.50 ml.

In reality it was a mere coincidence with the law that additivity was obtained when liquids were examined at their normal boiling points. But it was soon realised that molecular volume was not purely an additive property. Hence the use of molecular volume in deciding the chemical constitution is limited. Another property, related to molar volume, called *parachor*, has been found to be far more useful for this purpose, as discussed below.

MECLEOD'S RELATIONSHIP—THE PARACHOR

We know that surface tension is due to the inward force o the molecules and hence it is related to the structure of the molecules.

Macleod in 1923 pointed out that

$$\frac{\gamma^{1/4}}{D - d} = C \qquad ...(1)$$

where γ is surface tension of the liquid, 'D' its density and 'd' the density of its vapour at the temperature of experiment and 'C' is a constant. The equation holds good over a wide range of temperature. Multiplying both sides by M,

$$\frac{\mathrm{M}\gamma^{14}}{\mathrm{D} - \mathrm{d}} = \mathrm{MC} = [\mathrm{P}] \qquad ...(2)$$

where 'M' is the molecular weight of the liquid. The product M × C is called parachor and is represented as [P]. At ordinary temperature far away from the critical temperature 'd' is negligible as compared to D. Then Eq. (2) may be written as

$$\frac{\mathrm{M}\gamma^{1/4}}{\mathrm{D}} = [\mathrm{P}] \qquad ...(3)$$

But $$\frac{\mathrm{M}}{\mathrm{D}} = \text{molar volume}$$

$$\therefore \qquad [\mathrm{P}] = \text{molar volume} \times \gamma^{1/4} \qquad ...(4)$$

Thus parachor may be defined as the product of molar volume of a liquid and its surface tension raised to the power 1/4.

If the two liquids have same surface tension under a given set of conditions, their parachors are proportional to their molar volumes.

Suppose, at a particular temperature, $\gamma = 1$. Then, since M/D is molar volume, the equation (3) may be write as follows:

$$[\mathrm{P}] = \text{Molar volume}$$

Thus, the parachor may also be defined as *the molar volume of a liquid at a temperature at which its surface tension is unity.*

Parachor has been found to be more useful than molar volume in deciding between different alternative structures of compounds.

Sugden concluded that parachor is largely an additive property. This is supported by the following observations:

1. *The isomeric compounds of the same family* (such as esters, alcohols, etc.) are having almost the same parachor. For instance, there are six esters having the molecular formula $C_6H_{12}O_2$. Their

parachors, as included in Table 5 are very close to one another.

Table 5 : Parachors of Esters of Formula $C_6H_{12}O_2$

Ester	*Parachor*	*Ester*	*Parachor*
Methyl valerate	292.5	iso-Amyl formate	293.6
Ethyl isobutyrate	292.2	iso-Butyl acetate	295.1
Ethyl butyrate	293.6	n-Propyl propionate	295.3

2. *The difference between the parachors of successive numbers of different homologous series is nearly the same,* as show for hydrocarbons in Table. 5.

Table 6 : Parachors of paraffins

Paraffins		*Parachor*	*Difference*	*Parachor for one*
Name	*Formula*			*CH_2 group*
Ethane	C_2H_6	110.5		
			40.3	40.4
Propane	C_3H_8	150.8		
			119.3	39.8
Hexane	C_6H_{14}	270.1		
			39.2	39.2
Heptane	C_7H_{16}	309.3		
			35.7	35.7
Octane	C_8H_{18}	345.0		
			79.2	39.6
Decane	$C_{10}H_{12}$	424.2		

From the above table 6 it can be seen that the average difference corresponding to the parachor of the CH_2 group, from a similar study of several other homologous series (such as alcohols, ethers esters, aldehydes, ketones, etc.), has been found to be 39.0.

Atomic Parachors

The values for a given compound are expressed as sum of the numbers and nature of atoms present called *atomic parachors*. Knowing that each CH_2 group has a parachor value of 39, while decane $C_{10}H_{22}$ has a parachor value of 424.2, it follows that

since $C_{10}H_{22} = 10(H_2) + 2H$

$$\therefore \quad 424.2 = 10 \times 39 + 2 \times \text{atomic parachor of hydrogen}$$

$$\text{Atomic parachor of hydrogen} = \frac{424.2 - 390}{2} = 17.1$$

$$\text{Atomic parachor of carbon} = 39.0 - 34.2 = 4.8$$

If the atomic parachors of carbon and hydrogen are known it becomes possible to determine atomic parachors of other elements. For example atomic parachor of oxygen has been obtained from ethers, that of nitrogen from amines and those of chlorine, bromine and iodine from the corresponding halides.

Structural Parachors

It was soon reported that parachor, like molar volume, though large an additive property, has been partly a constitutive property as well.

Table 7 : Atomic and Structural Parachors

Atoms, Groups or Linkage	*[P]* Sugden (1924)	Mumford and Phillips (1929)	Vogel (1948)
C	4.8	9.2	8.6
H	17.1	15.4	15.7
O	20.0	20.0	19.8
	12.5	17.5	—
Cl	54.3	55.5	55.2
Br	68.0	69.0	68.8
I	90.0	90.0	90.3
$>C=O$	—	—	44.4
— OH	—	—	30.2
— COOH	—	—	73.7
— NO_2	—	—	73.8
Single bond	0	0	0
Double bond ($>C=C<$)	23.2	19.0	19.9
Triple bond ($-C\equiv C-$)	46.6	38.0	40.6
Rings 3-membered	16.7	12.5	12.3
4-membered	11.6	6.0	10.0
5-membered	8.5	3.0	4.6
6-membered	6.1	0.8	1.4

For example, the parachor of ethylene (C_2H_4), which should be = $2 \times 4.8 + 4 \times 17.1 = 78.0$ has been in reality, 99.5. Hence, the double bond should make a contribution of 21.5. Further careful determinations in the case of unsaturated compounds having ethylenic double bond have shown that each ethylenic double bond makes a contribution, on an average of 23.2. Similarly, triple bond has been making a contribution of 46.6.

The values of atomic and structural parachors for different atoms, linkages etc. have bee determined. The values for some important atoms and structural factors are give below with the help of these the parachor for a give compound is calculated o the basis of an assumed structure. If this value agrees with the value obtained from experimental determinations of surface tension, the structure is correct.

The values of the atomic and structural parachors computed by Sugden (1924) were revised by Mumford and Phillips (1929). Vogel (1948) revised these values further.

CALCULATION OF PARACHORS OF COMPOUNDS

The parachors of compounds can be calculated with the help of the above table. One example is demonstrated below:

Parachor of Acetone $\left(\begin{matrix} CH_3 \diagdown \\ \quad C{=}O \\ CH_3 \diagup \end{matrix}\right)$. It may be calculated as illustrated below:

$$3C = 3 \times 4.8 = 14.4$$

$$6H = 6 \times 17.1 = 102.6$$

$$1O = 1 \times 20.0 = 20.0$$

$$\text{1 double bond} = 1 \times 23.2 = 23.2$$

$$\text{Parachor of acetone} = 160.2 \quad (\text{Observed value} = 161.2).$$

Applications of Parachor in Deciding Structures

1. *Structure of quinone :* For quinone $C_6H_4O_2$, the following two structures have been proposed:

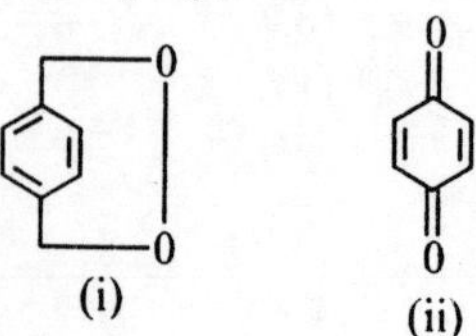

The observed value of parachor for quinone is 236.8, which of the two structure is correct?

(Parachors for H = 17.1, C = 4.8, O = 20.0, double bond = 23.2 and six membered ring = 6.1).

For structures (i) and (ii) we calculate the parachor values separate.

For (i) 6 Carbon atoms = 6 × 4.8 = 28.8

4 Hydrogen atoms = 4 × 17.1= 68.4

2 Oxygen atoms = 2 × 20 = 40

3 double bonds = 3 × 23.2= 69.6

2 six membered rings = 2 × 6.1 = 12.2

Total = 219.0

For (ii) 6 Carbon atoms = 6 × 4.8 = 28.8

4 Hydrogen atoms = 4 × 17.1= 68.4

2 Oxygen atoms = 2 × 20 = 40

4 Double bonds = 4 × 23.2= 92.8

1 six membered ring = 1 × 6.1 = 6.1

Total = 236.1

Since the observed value 236.8 is nearly the same as that calculated value of (*ii*), therefore the true structure of quinone is (ii).

2. *Structure of paraldehyde :* Acetaldehyde forms a liquid polme known as paraldehyde $(C_2H_4O)_2$. The two proposed structures for the polymer are:

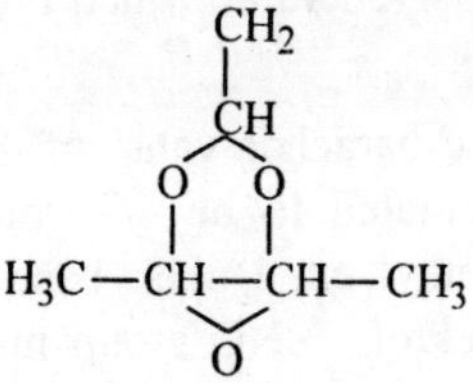

CH₂ | CH (bridging O O) ; $H_3C-CH-CH-CH_3$ with O bridge below

(A)

$CH_3CH(OH) . CH_3CH(OH) . CH_2CH{=}O$

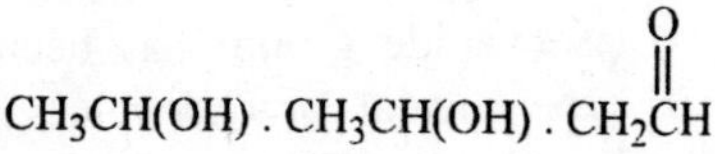

(B)

The value of parachor as calculated from its surface tension is 298.7. Which of the proposed structures is correct?

(Parachors of H = 17.1, C = 4.8, O = 20.0, six membered ring = 6.1 double bond = 23.2).

Solution:

We calculate prachors for (A) and (B) separately.

For (A) 6 Carbon atoms = 6 × 4.8 = 28.8

12 Hydrogen atoms = 12 × 17.1 = 205.2

3 Oxygen atoms = 3 × 20.0 = 60.0

1 six membered ring = 1 × 6.1 = 6.1

Total = = 300.1

For (B) 6 Carbon atoms = 6 × 4.8 = 28.8

12 Hydrogen atoms = 12 × 17.1 = 205.2

3 Oxygen atoms = 3 × 20.0 = 60.0

1 Double bond = 1 × 23.2 = 23.2

Total = = 317.2

Since the observed value 298.7 is the same as the calculated value for structure (A), therefore the correct structure of paraldehyde is:

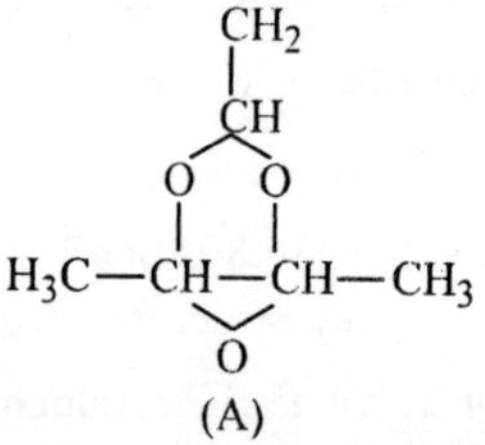

```
              CH2
              |
              CH
             /  \
            O    O
            |    |
   H3C—CH—CH—CH3
          \  /
           O
```

(A)

3. *Structure of benzene* : The calculated value of parachor taken fro Kekule formula agrees with observed value which is found to be 206.

4. *Structure of isocanide group* : The parachor value of —NC (isocyanide group) has been calculated to be 65 from the experimental parachor values for methyl, ethyl, phenyl and other isocyanides. The possible structures of —NC group may be given as:

—N=C I	and	$-\overset{+}{N}\overset{\rightarrow}{=}C$ II
Calculated parachor		Calculated parachor
= 12.5 + 4.8 + 23.2		= 12.5 + 4.8 + 46.6 – 1.6
= 40.5		= 62.3

The structure II has been preferred which has also been established by Raman spectra and dipole moment studies.

5. *Odd electro bound I :* The parachor value of odd electro bond has been found to be – 11.6 b Sugden. In the case of SF_6, if all the bonds are considered to be covalent ones then the calculated parachor of the molecule would be 198.2. The experimental value of SF_6 is 144 only. The difference has been explained on the fact that four of the fluorine atoms are attached with odd electron bonds.

   In this way odd electron bonds in various chelated compounds have been proposed.

6. *Deciding the nature of valency :* Parachor provides a valuable tool in deciding the nature of bonds in compounds of nitrogen, sulphur and phosphorus. For instance, the nitro group may be represented by the following three structures:

I	II	III
—N(=O)=O	—N(→O)=O	—N(–O–O–) (3-membered ring)
Parachor = 12.5+2×20 + 2 × 23.2 = 98.9	12.5+2×20+23.2 – 1.6 = 74.1	12.5+2×20+16.7 (for a 3 membered ring) = 69.2

The study of parachors in large number of nitro compounds reveals that the contribution of — NO_2 group is about 73.0. Therefore the structure II is most probable.

7. *Structure of benzil :* Two structures have been proposed for benzil.

I	II
$C_6H_5—C(=O)—C(=O)—C_6H_5$	$C_6H_5—C(–O)—C(–O)—C_6H_5$
Calculated parachor = 476	Calculated parachor = 464.4

The experimental value of benzil is 480.0. Hence structure I is most probable.

Parachor Anomalies

It must be pointed out that the use of the parachor for the solution of structural problem is not fee from objections. For example:

(i) Parachor measurements indicated that organic oxides, containing the — N_3 group, and aliphatic diazo-compounds have cyclic structure. It appears to be provided by electron diffraction measurements, however, that the groups are actually linear. The erroneous conclusions may perhaps have been due to the failure to take into account the possibility of resonance.

(ii) Serious parachor anomalies exist in several organometallic compounds: in the dialkyl sulphide derivatives of palladium chloride : having the general formula $(R_2S)_2$ $PcDCl_2$, the parachor equivalent of the palladium atom decreases from 3, when R is methyl, to – 7 when R is amyl.

(iii) Similar to (ii), a steady decrease in the contribution of the metal atom occurs in the mercury mercaptides. Hg $(SR)_2$, and in other series of compounds as the length of the hydrocarbon chain is increased.

In (ii) and (iii), it is probable that as the central atom becomes more completely covered with alkyl groups it becomes less and less able of making its contribution to forces of cohesion and to surface tension. In such circumstances the normal additivity relationships of the parachor might be expected to fail.

SOME SOLVED PROBLEM

Problem 1:

Calculate the molar refraction of CCl_4, if its refraction index, N_D^{293} is 1.4573 and the density at 293 K is 1.595 kg.dm^{-3}.

Solution:

$$R_M = \frac{\mu^2 - 1}{\mu^2 + 2} \cdot \frac{M}{D}$$

Substituting $\mu_D^{293} = 1.4573$

$$M = [12 + 4 \times 35.4]$$

$$= 153.6 \text{ gm} = 0.1536 \text{ kg}$$

$$d = 1.595 \text{ kg dm}^{-3}$$

$$R_M = \frac{(1.4537)^2 - 1}{(1.4573)^2 + 2} \cdot \frac{0.1536}{1.595}$$

$$= 26.24 \text{ cm}^3 \text{ mol}^{-1}.$$

4

DETERMINATION OF MACROMOLECULES

INTRODUCTION

It is known that the gas laws are based of experiments quite independent of any theory of the nature of gas. However a simple gas theory that *a gas consists of a very large number of molecules in rapid motion in a vacant space* can be very readily employed to derive and justify these laws mathematically. This theory is termed as the *Kinetic Molecular Theory of Gases.*

This is named kinetic molecular theory because any model for explaining the behaviour of gases must involve both the molecular concept and motion of the particles. This theory was developed by Bernoulli, Clausius, Kronig, Maxwell, Kelvin, Boltzman and others.

The Principle of Corresponding States

In 1881, van der Waals demonstrated that if the pressure volume and temperature of a gas are expressed in terms of its critical pressure critical volume and critical temperature, then an important generalization called the principle of corresponding states would be obtained.

According to van der Waals equation, we have

$$\left(P + \frac{a}{V^2}\right)(V - b) = RT \qquad ...(1)$$

This equation has constants 'a' and 'b' which cannot be measured directly. Therefore, this equation is not applicable to all gases and in order to make it so, these constants have to be removed.

It has been found experimentally that near the critical point all gases give the same type of isotherms as carbon dioxide, but the actual position of the critical isotherms gets varied with the critical constant of the gas studied. The similarity is great enough to postulate that if for each different gas a suitable scale of pressure, volume and temperature is selected, it will be possible to superimpose all the critical isotherms on one another. If this were done of the critical isotherms it becomes probable that all other isotherms would coincide.

Under such circumstances, there should be one isothermal diagram and one fundamental equation for all gases. Such a scale will be one in which the pressure, volume, and temperature will be expressed as fractions of the critical pressure, volume and temperature.

In order to obtain such an equation, the values of P, V and T must be expressed as corresponding fractions of P_c, V_c and T_c, the critical pressure, volume and temperature,

$$\pi = P/P_c \quad \text{or} \quad P = \pi P_c$$

$$\phi = V/V_c \quad \text{or} \quad V = \phi V_c$$

$$\theta = T/T_c \quad \text{or} \quad T = \theta T_c$$

where π, ϕ and θ are referred to as *reduced pressure, reduced volume and reduced temperature respectively.*

On substituting the values of P, V and T in the van der Waals equation for one mole of an ideal gas, we get

$$\left(\pi P_c + \frac{a}{\phi^2 V_c^2}\right)(\phi V_c - b) = R_q T_c \qquad ...(2)$$

$$P_c = \frac{a}{27b^2}, \quad V_c = 3b$$

$$\text{and} \quad T_c = \frac{8a}{27Rb}$$

On substituting these values in equation (2), we get,

$$\left(\pi \frac{a}{27b^2} + \frac{a}{9\phi^2 b^2}\right)(3\phi b - b) = R\theta \frac{8a}{27Rb}$$

or
$$\frac{a}{27b^2}\left(\pi + \frac{3}{\phi^2}\right) \times b(3\phi - 1) = \frac{a}{27b} 8\theta$$

or
$$\frac{a}{27b}\left(\pi + \frac{3}{\phi^2}\right)(3\phi - 1) = \frac{a}{27b} 8\theta$$

or $$\left(\pi + \frac{3}{\phi^2}\right)(3\phi - 1) = 8\theta$$

This equation involving neither of the characteristic constants a, b, T_c, P_c and V_c is known as van der Waals reduced equation of state. This equation is applicable to all substances (liquid or gaseous) irrespective of their nature, because it is not involving either of the characteristic constants.

It follows from the above equation that *fi two substances are at the same reduced temperature* (θ) *ad pressure* (π) *they must have the same reduced volume* (ϕ). This statement is called the law of corresponding states or principle of corresponding states.

Significance of Law of Corresponding States

It has a great significance in the study of the relationship between physical properties ad chemical constitution of various liquids. As it is known that the pressure has practically no effect on liquids, so while studying the relation between physical properties and chemical constitution the determination of these properties should be performed at the same reduced temperature.

It has been found that the *boiling point of a liquid at atmospheric pressure is almost 2/3rd of its critical temperature* (T_c) *when expressed in absolute scale.* Consequently, boiling point is considered as the *corresponding temperature.* So, various physical properties of liquids such as molecular volumes of various liquids should be compared at their respective boiling points.

In actual practice one cannot use the above reduced equation of state directly. One makes use of the graphs between compressibility factor 'Z' and the reduced pressure at different reduced temperatures. The same graph is applicable to all gases. This can be seen from the following considerations, since $Z = \frac{PV}{RT}$.

Writing this in terms of reduced terms,

We have $$Z = \frac{(\pi P_c)(\phi V_c)}{R(\theta T_c)} = \left(\frac{P_c V_c}{RT_c}\right)\frac{\pi\phi}{\theta} = \frac{3}{8}\frac{\pi\phi}{\theta}$$

Thus, if two gases are having the same reduced temperature and reduced pressure, then their reduced volume will be the same, thus R.H.S. is independent of the as, thus all gases will have the same value of 'Z'.

If isotherms are made to plot in terms of the reduced variables, the same curve should be obtained for all gases. The principle of corresponding states thereby enables the properties of real gases to be plotted.

Liquefaction of gases

For every known gas there is a temperature above which it is impossible to liquefy a gas, o matter how great the pressure may be. This temperature is known as the critical temperature. The minimum pressure necessary to liquefy a gas at its critical temperature is its critical pressure.

There are many gases such as ammonia, chlorine, sulphur dioxide, carbon dioxide, etc. which can be liquefied by the application of a suitable pressure alone because they are having high critical temperatures. On the other hand, there are gases such as hydrogen, oxygen, nitrogen and helium which cannot be liquefied by the simple application of a suitable pressure alone because they are having very low critical temperatures. These gases were regarded as *permanent* because the real significance of critical temperature was misunderstood for a longtime. But now it is known that these gases could be liquefied if they are first cooled below their respective critical temperatures and then made to compress.

The various methods used for the liquefaction of gases are as follows:

(1) *By the use of freezing mixture:* Faraday (1823) was able to liquefy a number of SO_2, Cl_2, CO_2, etc. by subjecting them to high pressures at low temperatures. He used a V-shaped tube, in one arm of which the gas was prepared and i the other end it liquefied under its own pressure and with the help of external cooling. But Faraday was not able to liquefy the gases like hydrogen, oxygen and nitrogen, even when the pressure was as high as 2790 atmospheres. These gases were, therefore, known as *permanent gases*.

In 1869, Andrews studied the effect of pressure on CO_2 at different temperatures and observed that for each gas there exists a temperature, called *critical temperature*, above which it is not possible to liquefy the gas by compression no matter how great may be the pressure. The ignorance of this fact was the main reason why the early attempts to liquefy the so-called 'permanent' gases got failed.

(2) *By the use of volatile refrigerants:* This method is based on the fact that when a volatile compound evaporates rapidly, it absorbs

its latest heat of vaporisation from the surroundings and consequently temperature of the surroundings becomes low.

Pictet was successful in liquefying CO_2 by the cooling caused by the evaporation of liquid SO_2 around it. Thilosier (1834) succeeded in solidifying liquid CO_2 by allowing it to evaporate freely under atmospheric pressure. Thilosier, obtained temperature – 100 to – 110°C by allowing a mixture of a solid CO_2 (dry ice) and ether (called Thilosier's mixture) to evaporate under reduced pressure.

Faraday was able to liquefy most of the common gases by using Thilosier's mixture. He however failed to liquefy O_2, N_2, H_2, etc. and the called them permanent gases.

(3) *By the use of Joule-Thomson effect* (By expansion of cold compressed gases).

JOULE-THOMSON EFFECT

The English physicists, James Joule and William Thomson (later Lord Kelvin), observed that if a *gas under high pressure is allowed to expand into a region of low pressure it suffers a fall in temperature.* This phenomenon is termed as Joule Thomson effect.

Joule-Thomson effect *provides further support to the view that attractive forces do exist between gas molecules.* When the gas expands, the molecules fall apart from one another. Therefore, work has to be done so as to overcome the cohesive or attractive forces which tend to hold the molecules together.

Hence, work is done by the system at the expense of the kinetic energy of the gaseous molecules. Consequently, the kinetic energy would decrease and since this is proportional to temperature, cooling results. It may be noted that in this case no external work is being done by the as in expansion.

Inversion Temperature

Experiment has revealed that gases become cooler during the Joule-Thomson's expansion *only if they are below a certain temperature called the* inversion temperature, T_i. The inversion temperature is characteristic of each gas. It is related to van der Waals constant a and b of the gas connected by the following relation:

$$T_i = \frac{2a}{Rb} \qquad ...(1)$$

At the inversion temperature, there occurs no Joule-Thomson effect. Thus, if a gas under pressure is allowed to pass through a porous plug and expands adiabatically into a region of very low pressure *at the inversion temperature,* there occurs neither fall no rise in temperature. If, however, the expansion occurs *above* the inversion temperature, there occurs a small *rise* of temperature and if it occurs *below* the inversion temperature there occurs a small *fall* of temperature.

In most gases, the inversion temperature lies within the range of ordinary temperature. Hence, they are cooled in Joule-Thomson's expansion. Hydrogen and helium, however, have *very low inversion temperatures*. Thus, at ordinary temperatures, these gases get *warmed* up instead of getting cooled in the Joule-Thomson's expansion. But if hydrogen is first cooled to – 80°C which is its inversion temperature and helium is fist cooled to – 240°C which is its inversion temperature, then these gases also get cooled on expansion in accordance to the Joule-Thomson effect.

Linde process of the liquefaction of gases, like air, is based on Joule-Thomson effect.

The actual steps of the process will be clear. Purified dry gas is first of all compressed to a pressure of 200 atmospheres. The compressed gas is then allowed to pass through coils C, cooled by water or ammonia. The compressed and cooled gas is allowed to pass through a long spiral type of tube T. After travelling through it the gas gets diffused through a valve into a chamber (D) where it expands suddenly. The pressure inside the chamber is 50 atmospheres The control valve V is adjusted from outside. The as is cooled ad is made to pass back to the compression over the incoming gas. Thus the incoming gas gets further cooled before expansion takes place inside the chamber. On repeating the cycle several times, the temperature of the expanding gas would go sufficiently down ad a part of it gets condensed. The liquid collects at the bottom of the chamber from where it is drawn off. The whole arrangement is insulated by wool or left to avoid heating of the tubes.

By Adiabatic Expansion

If a gas is allowed to expand against an external pressure (say against a piston of an engine), it does some work. If a system is closed and insulated so that neither heat could enter the system nor go out of it, then the expansion would be almost adiabatic. This work done against the external pressure would be at the expanse of kinetic energy of the gas molecule. Consequently, a fall in temperature takes place. This principal combined with Joule-Thomson effect finds use in Claude's process for the liquefaction of air.

Air free from CO_2 and moisture is made to compress to 40 - 50 atmospheres and then cooled by passing through a precooler. The gas then splits into two steams at B. A pat of the air enters the condenser C directly, where the other part enters the cylinder D where it undergoes expansion adiabatically, whereby the temperature of the gas falls. The cooled gas then enters the liquefying chamber via the side tube E.

As this cooled gas goes up in the chamber, it cools the air passing down the Jet J whereby further cooling takes place because of Joule-Thomson effect. The cooled gas then goes out through the precooler tube T, where it cools the incoming air. This cycle continues till sufficient low temperature gets reached and air would begin to liquefy. The liquefied air coolects at the bottom of the condenser.

The Claude's process has been observed to be much more economical than Linde's because a part of the work in compression can be recovered while in Linde's process work is carried out i free expansion and is lost. The work gained can be used to work the compressors.

Production of Low Temperatures by Adiabatic Deagnetization of a Paramagnetic Substance

It is possible to attain very low temperatures in the velocity of absolute zero by :

(i) Joule-Thomson expansion of hydrogen

(ii) Using liquid helium, and

(iii) By *making use of the magnetic properties of paramagnetic substances* such as the salts of the rare earths.

The para-magnetism of these substances is attributed to the presence of odd (unpaired) electrons.

The method of adiabatic demonetization of a paramagnetic substance was first of all suggested i 1926 by Debye and independently i 1927 by Giauque. This method is based on the fact that *the temperature of thermally isolated paramagnetic substance drop shwne jit gets demonetized.* This can be best understood by considering the following equation for a magnetic system:

$$dS = \frac{C_b dT}{T} + \varepsilon_0 V\left(\frac{\partial M}{\partial T}\right)_B dB \qquad ...(1)$$

where M refers to the magnetization in the presence of the external magnetic field B; V the volume, ε_0 the vacuum permeability and C_B the heat capacity of the system at constant magnetic field.

The magnetization of a paramagnetic substance at constant magnetic field B always decreases with increase in temperature T so that the partial derivative $(\partial M/\partial T)_B$ is always negative. Hence, when the substance is demagnetized reversibly ad adiabatically (so that dB is negative and dS = 0), we get

$$\frac{dT}{T} = -\frac{\varepsilon_0 V}{C_B}\left(\frac{\partial M}{\partial T}\right)_B dB \qquad ...(2)$$

or

$$\left(\frac{\partial T}{\partial B}\right)_S = \frac{\varepsilon_0 TV}{C_B}\left(\frac{\partial M}{\partial T}\right)_B \qquad ...(3)$$

Eq. (3) is known to form the basis of the principle of adiabatic demagnetization which gives the rate of change of temperature with magnetic field upon reversible adiabatic demagnetization. It can be seen from Eq. (2) that dT/T is negative so that dT is also negative. Hence a paramagnetic substance would cool upon adiabatic demagnetization.

The temperature-entropy (T-S) curve for a paramagnetic salt, such as gadolinium sulphate octahydrate, is depicted. From the curve it can be seen that in the unmagnetized state (when B = 0), the spins of the unpaired electrons of the paramagnetic material get oriented in a random manner in all directions. Thus, there occurs a disorder.

If, however, the substance is made to magnetize, the spins, and hence the magnetic moments associated with them, tend to align themselves with the magnetic field, thus reducing disorder. In other words, it means that the unmagnetized substance is having greater entropy than the magnetized substance at a given temperature. In the first stage of the process, the paramagnetic substance, in contact with a liquid helium bath at 1 K, would be *isothermally magnetized.* Under isothermal conditions (T = constant, dT = 0), Eq. (1) gets reduced to

$$TdS = \mu_0 TV(\partial M/\partial T)_2 dB \qquad ...(4)$$

As now dB is positive and $(\partial M/\partial T)_B$ is always negative, an increase in B causes an overflow of heat and hence a *decrease* in entropy. This is exhibited by the T = constant line on the T-S curve.

In the second stage, the substance is removed and *adiabatically demagnetized.* Now it cools. This is depicted by the S = constant line on the T-S curve. The extent of cooling has been found to depend upon the strength of the applied magnetic field, the intensity of magnetization and the heat capacity of the paramagnetic salt.

The experimental set-up used for carrying out adiabatic demagnetization. The paramagnetic substance A is made to suspend inside a cryostat which gets cooled by liquid helium boiling under reduced pressure. During the first isothermal stage, the space D is having helium gas; this space will permit heat exchange between the substance and liquid helium. During the second stage, the space D gets evacuate to establish adiabatic conditions. The cycle of isothermal magetization, followed by adiabatic demagnetization, gets repeated to get millidegree temperatures.

Even in the absence of external field B, which is usually of the order of 1-4 tesla, there exists a local magnetic field (or residual field) B_0 which is present in the paramagnetic substance which is arising mainly from magnetic interactions between the dipoles. B_0 is very small compared with B, its magnitude being of the order of 0.01 tesla. If T_1 refers to the initial temperature, then the final temperature T_2 would be given by the formula.

$$T_2/T_1 = B_0/B$$

If $T_1 = 1$ K,

$B_0 = 10^{-2}$ tesla, B = 1 tesla, then $T^2 \approx 10^{-2}$ K

We can remark here that the process of adiabatic demagnetization is also applicable to the magnetic moments of *nuclei*. The application of adiabatic demagnetization to nuclear magnetic moments has allowed a temperature of 10^{-6} K to be reached. However, one should remember that *it would not be possible to attain absolute zero of temperature by using a succession of adiabatic cooling processes*. This can be understood from the T-S curve. In 1949 W.F. Giauque won nobel prize in Chemistry for his contributions to thermodynamics and the production of low temperatures by adiabatic demagnetisation of a paramagnetic salt.

DETERMINATION OF MOLECULES

The molecular weight of a polymer depends upon the number of simple molecules joined together during polymerisation reactions, *i.e.,* upon the degree of polymerisation. But the polymerisation chains are broken at the different stages. Therefore, the final product will contain the macromolecules of different weights. For example all the molecules of styrene are similar and have the same mass. On the other hand, the masses of various individual molecules are different, *i.e.*, they are spread over a certain range. Hence it is necessary in their cases to take the average molecular weight. There are two types of average molecular weights:

Number-average Molecular Weight, M_n

It is obtained by dividing the total weight of the dispersed material with the number of molecules present, *i.e.*,

$$\overline{M_n} = \frac{n_1M_1 + n_2M_2 + ...}{n_1 + n_2 + ...}$$

where n_1, n_2, n_3, ... are the number of molecules having masses m_1, m_2, m_3, ... respectively.

The above equation may be written more conveniently as follows:

$$\overline{M_n} = \frac{\sum_i n_iM_i}{\sum_i n_i}$$

where n_i is the number of molecules of molecular weight M_i.

The molecular weight of the polymer obtained from measurements of colligative properties like osmotic pressure is the number-average molecular weight. This is due to the fact that osmotic pressure depends upon the number of particles and not upon the masses of all the particles.

Weight-average Molecular Weight, M_w

It gives representation to various molecular species in proportional to their weights in the give material. Hence, in the averaging process, the molecular weight of each individual species is multiplied by the weight and not by the number, *i.e.*,

$$\overline{M_n} = \frac{n_1M_1 + m_2M_2 + ...}{m_1 + m_2 + ...} \quad ...(a)$$

If n_1, n_2, ... denote the number of molecules having masses M_1, M_2, ..., then $m_1 = n_1M_1$, $m_2 = n_2M_2$, ... Hence the above equation may be written as follows:

$$\overline{M}_n = \frac{n_1M_1^2 + m_2M_2^2 + ...}{n_1M_1 + n_2M_2 + ...} = \frac{\Sigma n_iM_i^2}{\Sigma n_iM_i} \quad ...(b)$$

From Eqs. (a) and (b), it follows that $\overline{M}_w$ is always greater than $\overline{M}_n$.

In order to calculate the value of $\overline{M}_n$ from $\overline{M}_w$, it is necessary to know the relative numbers of the two species present in the polymer.

METHODS TO DETERMINE MOLECULAR WEIGHT

The following methods are generally employed for the molecular weight determination of macro-molecular or high polymers.

Sedimentation Method

It is the most modern and widely used method to determine the molecular weight of high polymers. This method was developed by T. Svedberg and his coworkers (1925-31).

Sedimentation is the fall of colloidal particles in a viscous medium under the influence of gravity. The rate of sedimentation of suspended particles under the influence of gravity is small. However, it can be increased considerably by subjecting it to ultracentrifuge technique. By this technique the force exerted on the particles becomes as great as million times the gravity. Suppose w is the angular velocity of rotation *i.e.,* rotational speed in radians per second.

Then, the force exerted on a particle at a point distant x from the axis of rotation would be w^2x/g times the gravity, where $w = 2\pi$ times the number of revolutions per second. Thus, it becomes possible to increase the rate of sedimentation by increasing number of revolutions per second.

Two different methods based on the use of ultracentrifuge have been devised.

Sedimentation Equilibrium Method

In this method, the solution taken in a column is rotated in a specially designed high speed instrument called ultracentrifuge. An equilibrium is said to be established under the influence of gravity in a sedimenting column when the amount of material driven outward by centrifugal force is exactly balanced by the amount diffused in the opposite direction due to Brownian and thermal motions.

Suppose particle is moving through a distance dx in time dt. Then under the influence of centrifugal forces the amount of substance (dw) which is moving outward per sq. cm would be cdx *i.e.*,

$$dw = c\ dx \qquad ...(1)$$

where c denotes the concentration of particles. We know that the diffusion coefficient (D) may be defined as the weight of material transferred across the plane of one square cm area when the concentration gradient (– dc/dt) is unity. Thus the value of dw may also be give as follows:

$$dw = -D\ .\ (dc/dx)dt \qquad ...(2)$$

When the sedimetation equilibrium is established, both the values of Eq. (1) and Eq. (2) becomes equal* but of opposite sign. Thus,

$$cdx = D\frac{dc}{dx}\ .\ dt$$

or

$$\frac{dc}{c} = \frac{1}{D}\cdot\frac{dx}{dt}\cdot dx \qquad ...(3)$$

We know,

$$D = \frac{RT}{6\pi nrN} \qquad ...(4)$$

and

$$\frac{dx}{dt} = 2r^2w^2x(\rho - \rho')/9\eta \qquad ....(5)$$

On substituting the equations (4) and (5) in Eq. (3), we get

$$\frac{dc}{c} = \frac{4}{3}\frac{\pi r^3 N w^2(\rho - \rho')}{RT} x\ dx \qquad ...(6)$$

But the volume of particles is $\frac{4}{3}\pi r^3 N$. If M is their molecular weight and ρ the density of particles, then

$$\frac{4}{3}\pi r^3 N = \frac{M}{\rho}$$

Thus, Eq. (6) becomes as follows:

$$\frac{dc}{c} = \frac{M\ .\ w^2(\rho - \rho')}{RT\rho} x\ dx \qquad ...(7)$$

On integrating the above equation between two concentration limits c_1 and c_2 and two depth limits x_1 and x_2, we get

$$\int_{c_1}^{c_2}\frac{dc}{c} = \int_{x_1}^{x_2}\frac{Mw^2(\rho - \rho')}{RT\rho} x\ dx \qquad ...(8)$$

or

$$\log_e \frac{c_2}{c_1} = \frac{Mw^2(\rho - \rho')(x_2^2 - x_1^2)}{2RT \cdot \rho}$$

$$= \frac{Mw^2(x_2^2 - x_1^2)}{2RT}\left(1 - \frac{\rho'}{\rho}\right) \quad ...(9)$$

Eq. (9) can be used for determining the size and molecular weight of particles.

Determination of Molecular Weight

It consists of a rotor A which revolves at a very high speed about the axis BB′. The solution under examination is taken in a cell C which is made of quartz windows. A beam of light passing through the solution is allowed to pass on a photographic plate P. The amount of light absorbed at two depths x_1 and x_2 can be measured by blackening the plate P.

If it is assumed the absorption of light is proportional to the concentration of particles, then the ratio of c_1 and c_2 can be calculated at two depth x_1 and x_2. The angular velocity w can be calculated by knowing the speed of the rotation of the ultra-centrifuge. If, the values of ρ and ρ', are known then we can calculate M, the molecular weight of the polymer from Eq. (9).

Limitation

This method takes several days for completion because low centrifugal forces of the order of 10,000 to 100,000 times the gravity are used to obtain accurate results. Under these conditions, the sedimenting column requires several days to achieve equilibrium.

Sedimentation Velocity Method

It is a quicker method which is based on the measurement of sedimentation velocity. This method is preferred because a much higher centrifugal force, even upto 500,000 times the gravity, may be used without introducing any appreciable error. This process is completed within a few hours.

Sedimentation is the fall of colloidal particles in a viscous medium under the influence of gravity. In sedimentation method, the solution is rotated in a specially designed high speed instrument called ultracentrifuge. As the colloidal particles undergo sedimentation on the rapid rotation of the solution, a moving boundary is formed, behind which only solvent is present. The beam of light, which is passing through the system and the falling on a photographic plate records the position of the boundary at various intervals of time. The following expression has bee used for calculating first the radius of the particles:

$$\frac{\ln x_2 / x_1}{w^2(t_2 - t_1)} = \frac{2R^2(\rho - \rho')}{9\eta} \quad ...(10)$$

where x_1 and x_2 are the distances of the boundary from the axis of rotation at times t_1 and t_2 respectively. η is the coefficient of viscosity of the medium and r is the radius of the particles. Eq. (10) may be derived as follows:

A particle undergoing sedimentation in a liquid column is subjected to a gravitational force, F, given by

$$F = (\text{Mass of particle} - \text{Buoyancy})g \quad ...(11)$$

where g is the Acceleration due to gravity. The buoyancy factor will be equal to the volume of the particles multiplied by the density of liquid medium.

If the particle is assumed to be spherical of radius r, then

$$F = \left(\frac{4}{3}\pi r^3\rho - \frac{4}{3}\pi r^3\rho'\right)g$$

$$= \frac{4}{3}\pi r^3(\rho - \rho')g \quad ...(12)$$

where ρ and ρ' are the densities of particle and of dispersion medium respectively. The particle undergoing sedimentation meets a frictional resistance R which is proportional to its velocity, dx/dt at any give instant and is given by

$$F = f\frac{dx}{dt} \quad ...(13)$$

where f is the frictional coefficient. For spherical particles, f is equal to $6\pi\eta r$, *i.e.*,

$$F = 6\pi\eta r \quad ...(14)$$

where η is the coefficient of viscosity of the liquid medium. On substituting Eq. (14) in Eq. (13), we have

$$R = 6\pi\eta r\frac{dx}{dt} \quad ...(15)$$

Due to the influence of the above two opposing forces, the particles attain a state of equilibrium, *i.e.*, zero Acceleration, and attains a uniform velocity. The two opposing forces R and F are in equilibrium. *i.e.*,

$$F = R$$

$$\frac{4}{3}\pi r^3(\rho - \rho')g = 6\pi\eta r \cdot \frac{dx}{dt}$$

or
$$\frac{dx}{dt} = \frac{2r^2(\rho - \rho')g}{9\eta} \qquad ...(16)$$

In the case of colloidal particles, r is small. Therefore, the rate of sedimentation (dx/dt) will also be very small. Svedberg et al. designed ultracentrifuge by which the rate of fall could be increased. If ω is the angular velocity of centrifuge and x is the distance from the axis of rotation, then

Acceleration due to centrifugal field = $\omega^2 x$ ...(17)

The force exerted on a particle in this manner can be increased million times that of gravity. Hence in Eq. (16), g may be replaced by w^2x giving

$$\frac{dx}{dt} = \frac{2r^2(\rho - \rho')w^2x}{9\eta}$$

or
$$\frac{dx}{x} = \frac{2r^2(\rho - \rho')w^2}{9\eta}dt \qquad ...(18)$$

As the particle undergoes sedimentation, the values of x and dx/dt go on changing continually.

If the particles are at distances x_1 after time t_1 and x_2 after time t_2 from the axis of rotation, then the Eq. (18) on integration within proper limits becomes as follows:

$$\int_{x_1}^{x_2} \frac{dx}{x} = \int_{t_1}^{t_2} \frac{2r^2(\rho - \rho')w^2}{9\eta} \cdot dt$$

or
$$\log_e \frac{x_2}{x_1} = \frac{2r^2w^2(\rho - \rho')}{9\eta} \cdot (t_2 - t_1) \qquad ...(19)$$

The quantity $\frac{\log_e x_2 / x_1}{w^2(t_2 - t_1)}$ is known as sedimentation coefficient, S, which may be expressed as follows:

$$\frac{\log_e x_2 / x_1}{w^2(t_2 - t_1)} = S = \frac{2r^2(\rho - \rho')}{9\eta} \qquad ...(20)$$

If the value of S is known, the value of r can be calculated. If the particles are assumed to be spherical, then the molecular weight M is given by the following relation:

$$M = \frac{4}{3}\pi r^3 N \cdot \rho \qquad ...(21)$$

where N is the Avogadro's number.

Both the sedimentation methods give the weight average molecular weight, $\overline{M}_w$.

If the value of S for successive time intervals is constant, the system is monodispersive, *i.e.*, it contains particles of uniform size. If the boundaries are distinct, then the system may be supposed to be polydispersive, *i.e.*, it contains particles of different sizes.

VISCOSITY DETERMINATION METHOD

It is a very convenient method for determining the molecular weight of macromolecules in solution. It has been observed that the presence of macromolecules raises the viscosity of the solvent due to inhomogenieties introduced by large molecules.

If η_0 is the viscosity of pure solvent and η that of solution of a given macromolecule, then $\frac{\eta}{\eta_0}$ is known as the relative viscosity (η_r) of the solution, *i.e.*,

$$\eta = \frac{\eta}{\eta_0} \qquad ...(22)$$

Specific viscosity (η_{sp}) may be defied as the relative increase in viscosity and is given by

$$\eta_{sp} = \frac{\eta - \eta_0}{\eta_0} = \frac{\eta}{\eta_0} - 1 = \eta_r - 1 \qquad ...(23)$$

The ratio η_{sp}/c, *i.e.*, the relative increase in specific viscosity per unit concentration of macromolecule is known as reduced viscosity. The reduced viscosity depends upon the molecular weight of macromolecule. As reduced viscosity is to independent of concentration, it becomes necessary to extrapolate a plot of η_{sp}/c against c to zero concentration. This extrapolated value is known as the intrinsic viscosity (η_i). This may be expressed by the following expression:

$$\eta_i = \lim_{c \to 0}\left(\frac{\eta_{sp}}{c}\right) \qquad ...(24)$$

where c is the concentration of the macromolecule in grams per 100 ml.

In order go get intrinsic viscosity of a macromolecule, it becomes necessary to determine specific viscosity of the solution of macromolecule of different concentrations in a given solvent.

According to Staudinger, the intrinsic viscosity of macromolecule is related to the molecular weight (M) by the following relation:

$$\eta_i = kM^a \qquad ...(25)$$

where k and a are constants for specific macromolecule in a specific solvent.

If k and a are known for a macromolecule-solvent combination, the molecular weight of the high polymer is determined simply by knowing the value of intrinsic viscosity.

Determination of k and a

On taking logarithms of Eq. (25), we get

$$\log_e \eta_i = \log_e k + a \log_e M \qquad ...(26)$$

It is possible to determine the values of the constants k and a by measuring intrinsic viscosities of a series of samples of a given macromolecule substance, the molecular weight of which bee measured by other methods based on measurement of osmotic pressure, sedimentation coefficient, etc. The values of $\log_e \eta_i$ are the plotted against $\log_e M$ which according to Eq. (26) will be a straight line. The intercept will be equal to $\log_e k$ and the slope of this straight line will be equal to a.

Viscosity is a relative method because the molecular weight obtained by this method depends upon the molecular weight by some other method.

This method is most convenient because once the values of k and a are known for a give type of macromolecule and solvent system, all that is required is to measure the viscosities of solutions of known concentrations in the same solvent.

Osmotic Pressure Method

The molecular weights of macromolecules are very high, generally lying in the range above 10,000. Thus the depressions in freezing points or elevation in the boiling points of their solutions would be too small to be measured accurately. But the osmotic pressures of such solutions would be too small to be measured accurately. But the osmotic pressure of such solutions would be too small to be measured accurately. But the osmotic pressure of such solvents would be too high which can be measured with a far better degree of accuracy. Another advantage of osmotic pressure method is that the presence of a slight trace of an impurity of low molecular weight results in a serious error in freezing point or boiling point elevation whereas it would pass through the cellophone membrane and, therefore, would not alter the pressure measured.

The van't Hoff's equation is $P = \frac{cRT}{M}$ ...(1)

where P is the osmotic pressure of a solution of concentration c gm per litre, M is the molecular weight of the solute, R is the gas constant and T is the absolute temperature.

Eq. (1) is applicable in ordinary solution having 1% of the solute. However, it fails in the case of high polymers when the concentration is considerably less. The reason for this failure is that the long chain macromolecules enclose within their folds a large number of small molecules of solvent which are thus removed from the solution. Due to this, the effective concentration of the solution is much increased. Therefore, the van't Hoff equation is treated as a limiting law in the solutions of macromolecules and may be written as

$$\lim_{c \to 0}\left(\frac{P}{c}\right) = \frac{RT}{M} \qquad ...(2)$$

From Eq. (2) it is evident that the limiting value of P/c at zero concentration (*i.e.*, at infinite dilution) becomes equal to RT/M.

In the actual practice, the osmotic pressure (P) of different small concentration (c) are determined. The values of P/c are plotted against c. The intercept corresponding to c = 0 yields RT/M from which the molecular weigh M can be evaluated because R and T are known.

The apparatus used in measuring osmotic pressure of macromolecule solution consists of two capillary tubes fitted in a stainless steel chamber across which is clamped a semipermeable membrane of cellophane. The tube in which the liquid rises is very narrow. This is made to avoid excess entry of water which may cause the dilution of the solution. The chamber is filled with the solution of the macromolecule till the level project is the lower end of the capillary. It is then fixed in a large tube having the solvent. The osmotic pressure is determined by the rise in level of the solution in the capillary.

In the case of colloidal electrolytes, the osmotic pressure has bee found to be greater than the theoretical value. The reason for this is the ions pass through the membrane and are not distributed equally o both side of membrane. This effect is known as *Donnan membrane equilibrium*. This may be reduced by increasing the concentration of the colloidal electrolytes.

Diffusion Method

This method is also used in determining the molecular weight. Stokes-Einstin expression may be written as,

$$M = \frac{DR^3T^3}{162\pi^2N^2\eta^2\Delta}$$

Assuming the particles to be spherical, and where, M is the molecular weight of the dispersed phase, D is the dielectric constant of the medium, η is the viscosity of the solvent and Δ is the diffusion.

Thovert's equation (1942) can also be used, provided the particles are comparatively smaller. It is given by,

$$M = k/\Delta^2$$

If Δ refers 1 sq. cm per sec, then k is equal to 60×10^{-10}. For an As_2S_3 sol, the value of molecular weight comes out to be approximately equal to 6,0000.

LIGHT SCATTERING METHOD

If a strong beam of light is passed through vacuum, it is transmitted in the direction of travel and the path of the light cannot be detected. However, when light is passed through a medium containing discrete particles, it gets scattered in all directions in space and this is the cause of the turbid appearance of most of the colloidal solution.

This fraction of the incident light scattered per unit length of the solution through which it passes is called *turbidity* (τ), which is given by the expression

$$I_1 = I_0c^{-\tau l} \quad ...(1)$$

where I_0 is the intensity of the incident light beam ad I_1 that of the transmitted light beam after passing through a length l of the solution. In case of solution of most of the proteins and high polymers, turbidity is very small and it is determined by measuring the intensity of the light scattered at 90° to the beam. This can be done accurately by means of a simple type of photometer. A detecting photo cell is outed on a rotating arm to permit measurement of the light scattered at several angles, and fitted with a polaroid for observing the polarisation of the scattered light. Debye derived the relation between the turbidity and the molecular weight of the macromolecules, which is given by

$$M = \frac{1}{\lim_{c \to 0} H \frac{c}{\tau}} \qquad \text{...(2)}$$

where H is constant for specified colloidal dispersion, and is a function variation of the refractive index with concentration and the wavelength of the light. The turbidity and refractive index at different concentrations are measured and the quantity H (c/τ) is plotted against concentration when a lie a graph is obtained. The intercept gives the value of 1/M. Since the amount of light scattered by solution is very small, every care should be taken to see that the solution is free from impurities such as dust particles which would themselves scatter light considerably and introduce serious errors.

It is see from equation (2) that the turbidity of a solution of a give concentration (g . ml^{-1}) increases with increase in molecular weight of the polymer in contrast to osmotic pressure which decreases with increase in molecular weight.

Therefore this method gives accurate results for macromolecules having molecular weight more than 10^6 for which the osmotic pressure method becomes inaccurate.

Limitation

(i) As the amount of light scattered by solutions of macromolecules is very small, it becomes essential to free them of dust particles which would themselves scatter light considerably.

(ii) As the turbidity of solution of macromolecule of a give concentration (grams per l) increases with increase in molecular weight, it is evident that the light scattering method is more accurate in the case of polymers of very high molecular weights particularly above 1,000,000 where osmotic pressure method gives inaccurate results.

Intermolecular Forces or Van der Waals forces

It is now well established that forces of attraction exist between *neutral molecules* as well. These are termed as intermolecular forces or cohesive forces or van der Waals forces. These forces arise from three in types of interactions, viz.,

(1) *Dipole-Dipole Interaction*

(2) *Dipole-Induced Dipole Interaction, and*

(3) *Induced Dipole-Induced Dipole Interaction.*

1. Dipole—Dipole Interaction

The *polar molecules* which, though neutral, are having permanent dipoles. In such molecules the van der Waals forces are mainly attributed to electrical interaction between the dipoles known as *dipole-dipole interaction*. For instance, gases like ammonia, sulphur dioxide, hydrogen fluoride, hydrogen chloride, etc., are having permanent dipoles as a result of which appreciable dipole-dipole interaction exists between the molecules of these cases.

The magnitude of this type of interacting has been found to depend upon the dipole moment of the molecule concerned. The greater the dipole moment the greater would be the dipole-dipole interaction. Due to the attractive iteraction, these gases can be readily liquefied.

The average interaction energy of the two molecules with permanent dipole moments μ_1 and μ_2 is given by the expression

Interaction energy.

$$\phi(r) = -2\left(\frac{\mu_1\mu_2}{4\pi\varepsilon_0}\right)^2\left(\frac{1}{6r}\right)\left(\frac{1}{3kT}\right) \qquad ...(1)$$

where r refers to their separation, k Boltzmann constant and $4\pi\varepsilon_0$ the permittivity factor of the vacuum.

2. Dipole—Induced Dipole Interaction

A polar molecule may some times be able to polarise a neutral molecule which lies in its vicinity and thus brings about dipolarity i that molecule just as a magnet is able to induce magnetic dipolarity i a neutral piece of iron lying close by. The induced dipole then undergoes interaction with the dipole moment of the first molecule and thereby the two molecules get attracted together. The magnitude of this interaction would, evidently, depend upon the magnitude of the dipole moment of the polar molecule and the polarizability of the neutral molecule. The average interaction energy for the two molecules is given by the following expression:

$$\text{Interaction energy, } \phi(r) = -\frac{\mu_1^2\alpha_2}{(4\pi\varepsilon_0)^2 r^6} \qquad ...(2)$$

where μ_1 refers to the permanent dipole moment of molecule 1, α_2 refers the polarizability of molecule 2 and r, as usual, refers to the separation of the molecules. $4\pi\varepsilon_0$ as mentioned above, refers to the permittivity factor.

3. Induced Dipole—Induced Dipole Interaction. London Forces or Dispersive Forces

The van der Waals forces are also known to exist even in non-polar molecules like O_2 ad N_2 and also in non-polar monoatomic molecules like He, Ne, Ar, tc. This attraction has been evident from the condensation of these gases into liquids at sufficiently high pressures and low temperatures. This could not be understood for may years.

In 1930, F. London gave a satisfactory explanation for the existence of forces of attraction between *no-polar* molecules. This explanation is based on quantum mechanics. According to this view, electrons of a neutral molecule keep on oscillating with respect to the nuclei of the atoms. Because of this, at a given instant, positive charge will be concentrated in one region and negative charge in another region of the same molecule. Hence, a non-polar molecule would become momentarily self polarised. This polarised molecule maybe able to induce a dipole moment in a neighbouring molecule with antiparallel orientation.

The electrostatic forces of attraction between induced dipoles and the original dipoles (due to electron oscillation) are termed as London forces. These forces are also termed as dispersive forces because the well known phenomenon of dispersion of light is also related to these dipoles.

For a pair of adjacent molecules, London forces have been found to vary inversely as the *seventh power* of the distance between them, *i.e.*, $F \propto 1/r^7$. The magnitude of these forces, evidently, would be very appreciably small.

The van der Waals attraction is *non-polar molecules is therefore exclusively ascribed* to London forces. In polar molecules having permanent dipoles, however, other electrostatic forces because of dipole-dipole interaction also contribute towards the van der Waals attraction.

EQUATIONS OF STATE FOR REAL GASES

A number of equations of state have been given to describe the P-V-T relationship in real gases. The earliest ad the best known equation is the one which is given by van der Waals.

THE VAN DER WAALS EQUATION OF STATE

In 1873, van der Waals gave his famous equation of state for a non-ideal (*i.e.*, imperfect gas). He modified the ideal gas equation by postulating that the gas molecules were not mass points but behaved like rigid spheres having a certain diameter and that there existed *intermolecular forces of attraction between them*. The two correction terms introduced by van der Waals have been described below.

1. *Correction due to volume of gas molecules:* It is known that the derivation of the ideal gas equation PV = nRT is based on the assumption that the gas molecules are mass points, *i.e.*, they do not have any finite volume. van der Waals discarded this assumption and suggested that a correction term *nb* should be subtracted from the total volume V so as to get the *ideal volume which is compressible.*

In order to understand the meaning of the correction term b, consider two gas molecules as unpentrable and incompressible spheres, each of which is having a diameterσ. It can be seen that the centres of the two spheres cannot approach each other more closely than the distance σ.

For this pair of molecules, therefore, a sphere of radius σ and therefore of volume $\frac{4}{3}\pi\sigma^3$ constitutes what is termed as the excluded volume. The excluded volume per molecule would be therefore half the above, volume, viz., $\frac{2}{3}\pi\sigma^3$.

The actual volume of one gas molecule of radius

$$\text{r is } \frac{4}{3}\pi\sigma^3 = \frac{4}{3}\pi(\sigma/2)^3 = \frac{1}{6}\pi\sigma^3$$

$\therefore$ Excluded volume per molecule $= \frac{2}{3}\pi\sigma^3 = 4 \times \frac{1}{6}\pi\sigma^3$ = 4 times the actual volume oft he gas molecule.

The excluded volume would be *thus 4 times the actual volume of the molecule.* Evident, the excluded volume per mol of the gas would be $N_A \times 4 \times \frac{4}{3}\pi r^3 = b$ where N_A refers to the Avogadro's number. The compressible volume per mole of the gas would thus be V – b. If volume V of the gas is having n moles, the excluded volume would be nb. Thus, the ideal volume which is compressible is V-nb, the volume b per mole is also termed as co-volume.

2. *Correction due to intermolecular forces of attraction:* In the

derivation of the ideal gas equation, an assumption has been made that there are no intermolecular forces of attraction, we will consider a molecule which is lying somewhere in the midst of the vessel, as shown at the point A. Thus molecule is being attracted uniformly on all sides by the neighbouring molecules. These forces are able to neutralize one another ad there occurs no resultant attractive force on the molecule. However as the molecule approaches the wall of the vessel, as shown at B, it would experience attractive forces form the bulk of the molecules behind it.

Hence, it means that it will strike the wall with a lower velocity and will exert a lower pressure than it would have done if there was no force of attraction. It is, therefore, necessary to add a certain quantity to the pressure of the gas i order to obtain the *ideal pressure*. The correct pressure, therefore, world be P + p.

CALCULATION OF CORRECTION FACTOR P

The force of attraction exerted on a single molecule which is about to strike the wall evidently is dependent upon the *number of molecules per unit volume in the bulk of the gas*, *i.e.*, directly upon the *density* of the gas. Further, the number of molecules striking the wall at any given instant is also dependent directly upon the *density* of the as. Thus, the total inward attractive pull on the molecules which gives a measure of the correction factor, p, is proportional to the square of the density (ρ) of the gas, *i.e.*,

$$p \propto \rho^2 \qquad \text{...(2)}$$

But density is inversely proportional to the volume and if V refers to the volume occupied by one mole of a as, the value of p for one mole of a gas will be inversely proportional to the square of volume.

Hence $\qquad p \propto 1/V^2$

or $\qquad p = a/V^2 \qquad$ ...(3)

where a refers to a *constant* depending upon the nature of the gas.

The kinetic gas equation for *one mole* of a real gas, therefore, takes the following form

$$(P + a/V_m^2)(V_m - b) = RT \qquad \text{...(4)}$$

where V_m refers to the *molar volume* of the gas.

This is termed as van der Waals equation. The constant a and b are termed as van der Waals constants. These are characteristic of each gas.

Eq. (4) has been found to be valid for one mole of a gas only. If there are moles of a gas occupying volume V, then, the excluded volume will be given by b and the compressible volume, therefore, will be V – nb. The pressure correction factor p for n moles, in the light of Eq. 2. will be proportional to $n^2\rho^2$, *i.e.*

$$p \propto n^2\rho^2 \ \mu \ n^2 \times 1/V^2$$

Thus, $$p = an^2/V^2 \qquad ...(5)$$

Thus, the van der Waals equation for n moles of a as may be put as follows:

$$\left(P + \frac{n^2 a}{V^2}\right)(V - nb) = mRT \qquad ...(6)$$

Eq. 6 has been found to be more accurate than the ideal as equation PV = nRT for expressing the PVT behaviour of real gases. Thus, if one mol of carbondioxide is taken at 47°C and compress it to different pressures, the volume, as observed by experiment, is found to be closer to that calculated from van der Waals equation that to that calculated from the ideal gas equation. The departure from ideal gas equation would become more and more wide as the pressure continues to rise (Table 1).

Table 1 : Volume of Carbon Dioxide at 47°C under Different Pressures.

Pressure (atm)	*Volume (dm^3) of the gas*		
	Observed experimentally	*Calculated from van der Waals equation*	*Calculated from ideal gas equation*
1	27.20	26.20	26.30
10	2.52	2.53	2.63
40	0.54	0.55	0.66
100	0.098	0.10	0.20

UNITS FOR VAN DER WAALS CONSTANTS

The unit for the van der Waals constants a and b would depend upon the units in which *P* and *V* are expressed. It follows from Eq. 5 that the constant a is expressed by pV^2/n^2, *i.e.,* pressure × (volume)^2/mol^2. If pressure is expressed in atmospheres and volume in dm^3 the value of a will be in atm dm^6 mol^{-2}. As regards b, it is incompressible volume per mole, of a gas.

Hence it must have the same units as volume per mole, *e.g.*, dm^3 mol^{-1}.

The van der Waals constants for some gases have been included in Table 2.

The constant a may be regarded as *measure of the van der Waals forces of cohesion existing* between the molecules of a given gas. The greater the value of a, the greater is the strength of the van der Waals forces. The values of a for hydrogen and helium are very small, being 0.024 and 0.034 respectively, indicating that the cohesive forces (van der Waals forces) in these gases are very weak.

Table 2 : van der Walls constants for some common gas.

Gas	*a* (dm^6 *atm.* mol^{-2})	*b* (dm^3 mol^{-1})
Ammonia	4.17	0.0371
Argon	1.35	0.0322
Carbon dioxide	3.59	0.0427
Carbon monoxide	1.49	0.0399
Chlorine	6.49	0.0562
Ethane	5.49	0.0638
Ethylene	4.47	0.0571
Helium	0.034	0.0237
Hydrogen	0.024	0.0266
Hydrogen chloride	3.67	0.0408
Hydrogen bromide	4.45	0.0433
Methane	2.25	0.0428
Neon	0.21	0.0171
Nitric oxide	1.34	0.0279
Nitrogen	1.39	0.0319
Oxygen	1.36	0.0318
Sulphur dioxide	6.71	0.0564
Water	5.46	0.0305

The values of a for ammonia, carbon dioxide, chlorine and sulphur dioxide are very high being 4.17, 3.59, 6.49 and 6.71, respectively. This reveals that van der Waals forces in these gases are very strong. The greater the value of a, the greater would be the ease with which a gas can be liquefied.

Postulates of the Kinetic Molecular Theory of Gases

Kinetic theory of gases is based upon the following assumptions:

(*i*) *All gases are made up of very large umber of minute particles called molecules.*

(*ii*) The molecules are small and separated from one another by large distances so that the actual volume occupied by the molecules is almost negligible as compared to the total volume of the gas.

(*iii*) The molecules are always moving in all directions in straight lines with very high velocities. They keep on colliding against each other (molecular collisions) and against the walls of the vessel at very small intervals of time.

(*iv*) The molecules are spherical and perfectly elastic, and there occurs o loss of kinetic energy due to mutual collisions or collisions with the walls of the container. The kinetic energy may be transferred from one molecule to another but it is not converted into any other form of energy such as heat.

(*v*) The pressure exerted by a gas on the walls of the containing vessel is due to the bombardment of the molecules of the gas on the walls of the containing vessel.

(*vi*) The average kinetic energy of the molecules of gas is directly proportional to the absolute temperature.

(*vii*) There are no attractive forces between molecules or between molecules and the walls of the vessel in which the gas is contained. The molecules move completely independent of one another.

(*viii*) The motion imparted to the molecules by gravity is negligible in comparison to the effect of the continued collisions between them.

(*ix*) The laws of classical mechanics (in particular the Newton's second law of option) are applicable to gaseous molecules.

Justification for the Postulates

The above postulates are sufficiently justified by the following experiences:

1. If a gas is contained in a closed vessel, "it is expected that moving molecules will keep on hitting the walls of the vessel continuously

thus exerting pressure."

2. Molecules are very close in a compressed gas and intermolecular attraction should be high but this attraction must be negligible because even a highly compressed gas expands freely and fills the whole available space.
3. Gases intermix freely with other gases in all proportions. Even a heavier gas rises up to intermix with a lighter gas. This indicates that the effect of gravity is negligible i comparison to the effect of the continued collisions between them.
4. On heating, the temperature of gas rises. Heat energy supplied increases the speed of the molecules and their average kinetic energy increases thereby causing the rise of temperature.
5. It is known that gases can be compressed by the application of pressure and can also expand indefinitely filling the whole vessel no matter how large it may be. This reveals that gases are made of very large umber of minute particles separated from each other by large distances. Further these particles occupy very small volumes as compared to the total volume of the gas. In a gas like oxygen, hydrogen or nitrogen, at N.T.P. the volume occupied by the molecules themselves is only 0.014 percent of the total volume. The remaining 99.986 per cent of the volume is merely an empty space.
6. Dust particles i air are found to be in constant irregular zig-zag motion termed *Brownian movement*. This is an evidence for the very fast straight line motion of the molecules.[Postulate (*iii*)]
7. Rapid motion of a large number of molecules must involve frequent collisions between then and these collisions must be elastic as otherwise there will occur constant loss of kinetic energy and the molecules must come to stand still which does not happen.

DERIVATION OF THE KINETIC EQUATION FOR GASES

BY taking the postulates of kinetic theory into consideration, it is possible by applying the laws of classical mechanics to derive a mathematical expression which can explain the various gas laws. This expression is known as the *Kinetic Gas Equation*. Suppose a definite mass of a gas is contained in a cobical vessel and suppose that the length of each side of the cube = l cm

the total number of molecules = N

the mass of one molecule = m

The selection of the cubical shape of the container is purely a matter of convenience and involves no loss of generality because the pressure of a gas is independent of the shape of the container.

It is regarded that the container are smooth and get oriented perpendicular to the x, y and z-axes of a Cartesian coordinate system. The motion of molecules in the container, at any instant, is totally random.

Although, the molecules are moving in every possible direction the average velocity v of a molecule at any moment can be resolved into three directions X, Y, Z at right angles to one another.

Suppose a single molecule is moving with a velocity v.

Suppose v_x, v_y and v_z be the components of velocity v in the x, y ad z directions parallel to the three edges of the cube.

These three components get related to the resultant velocity v by the expression give below:

$$v = v_x^2 + v_y^2 + v_z^2 \qquad ...(1)$$

Suppose the motion of this molecule occurs between the two opposite faces A and B parallel to the x-axis. It strikes the face A at regular intervals.

The velocity of the molecule before it strikes the face A is v_x. But the molecule is perfectly elastic. Therefore it will rebound with the same velocity with the sign changed, *i.e.*, $-v_x$.

$$= 2mv_x \text{ gm. cm/sec} \qquad ...(2)$$

In order to strike the same face (A) against, the molecule will have to move to opposite face 'B' and come back. It means that the particle has to travel a distance 2l cm for each successive collision on the same face.

Thus, by traversing 2l cm., one impact is produced. By traversing 1 cm., 1/2l impacts get produced.

$\therefore$ By traversing v_x cm, $\frac{v_x}{2l}$ impacts get produced per second.

(v_x is velocity of the molecule that is the distance travelled i one second.)

Hence the number of impacts made by one molecule in one second

$$= \frac{v_x}{2l}$$

$\therefore$ Total change ion momentum per second (due to $\frac{v_x}{2l}$ impact) on face A

$$= 2mv_x \times \frac{v_x}{2l}$$

$$= \frac{mv_x^2}{l} \text{ gm cm/sec}^2$$

Total change in momentum per second because of the impacts of the molecule on two opposite faces A and B, along x-axis is therefore, double, *i.e.*, equal to

$$\frac{mv_x^2}{l} + \frac{mv_x^2}{l} = \frac{2mv_x^2}{l} \quad ...(3)$$

Also the total change of momentum per second due to impact of a single molecule on the two opposite faces along y-axis and two opposite faces along z-axis will be give as follows:

$$\frac{2mv_y^2}{l}, \frac{2mv_z^2}{l} \text{ respectively.}$$

Therefore, the total change in momentum per second o all the six faces of the cube, is equal to

$$= \frac{2mv_x^2}{l} + \frac{2mv_y^2}{l} + \frac{2mv_z^2}{l}$$

or $$= \frac{2m}{l}(v_x^2 + v_y^2 + v_z^2)$$

or $$= \frac{2mv^2}{l} \text{ gm cm/sec}^2 \quad ...(4)$$

$$(\because \; v_x^2 + v_y^2 + v_z^2 = v^2)$$

Thus total change of momentum, per second due to one molecule

$$= \frac{2mv^2}{l} \text{ gm cm/sec}^2$$

According to Newton's second law of motion, the rate of momentum is force *i.e.*,

$$\therefore \qquad \text{Force} = \frac{2mv^2}{l} \text{ dynes}$$

The pressure 'P' (force per unit area) is the total force divided by area of six faces of the cube ($6l^2$), *i.e.*,

$$P = \frac{\text{Force}}{6l^2} = \frac{2mv^2}{l} \times \frac{1}{6l^2} = \frac{1}{3}\frac{mv^2}{l^3}$$

or

$$P = \frac{1}{3}\frac{mv^2}{l^3} \qquad ...(5)$$

But l^3 is the volume of the cube 'V'

$$P = \frac{1\, mv^2}{3\, V}$$

Hence Pressure exerted by one molecule is

$$P = \frac{1}{3V} mv^2 \qquad ...(6)$$

The total pressure exerted by N different molecules of same mass ad having different velocities v_1, v_2, v_3 ... v_N will be as follows:

$$P = P_1 + P_2 + P_3 + ... + P_N$$

$$= \frac{1}{3V} mv_1^2 + \frac{1}{3V} mv_2^2 + \frac{1}{3V} mv_3^2 + ... + \frac{1}{3V} mv_N^2$$

[from Equation ... (6)]

$$P = \frac{1}{3V} m (v_1^2 + v_2^2 + v_3^2 + ... + v_N^2) \qquad ...(7)$$

If we multiply and divide the right had side of equation (7) by , we get

$$P = \frac{mN}{3V} \frac{(v_1^2 + v_2^2 + v_3^2 + ... v_N^2)}{N}$$

$$= \frac{mN}{3V} \times (\text{mean square velocity})$$

or

$$P = \frac{mN}{3V} \times (\text{root mean square velocity})^2$$

$$= \frac{mNc^2}{3V}$$

where N refers to the total number of molecules and $\sqrt{\bar{c}^2}$, the root mean square velocity give by

$$\sqrt{\bar{c}^2} = \sqrt{\frac{v_1^2 + v_2^2 + v_3^2 + \dots v_N^2}{N}}$$

$$PV = \frac{1}{3} mN\bar{c}^2 \quad \text{...(8)}$$

This is known as the Kinetic Gas Equation

It may be expressed as $PV = \frac{1}{3} M\bar{c}^2$ where M is the total mass of the gas.

or $$P = \frac{1}{3}\frac{m}{V}\bar{c}^2 = \frac{1}{3} D\bar{c}^2$$

where D is density of the gas.

KINETIC ENERGY AND TEMPERATURE

Translational kinetic energy of a body of mass m with velocity $\sqrt{\bar{c}^2}$ is give by as follows:

$$K.E. = \frac{1}{2} m\bar{c}^2$$

As the velocity of a molecule increases with the rise in temperature, it becomes possible to define mean kinetic energy in terms of temperature of the gas.

Suppose one mole of an ideal gas is under consideration. Then the total number of molecules is N (Avogadro's number).

$\therefore$ Kinetic energy per mol of the gas for N particles is given by

$$K.E. = \frac{1}{2} mN\bar{c}^2 \quad \text{...(9)}$$

The kinetic gas equation (8) for N molecules of a gas is

$$PV = \frac{1}{2} mN\bar{c}^2$$

or $$PV = \frac{2}{3} \times \frac{1}{2} mN\bar{c}^2$$

or $$= \frac{2}{3} K.E. \qquad (\because \frac{1}{2} mNc^2 \text{ is the kinetic energy})$$

On comparing this equation with ideal gas equation PV = RT for 1 mole of the gas, we obtain

$$RT = \frac{2}{3} K.E.$$

or $$K.E. = \frac{3}{2} RT$$

or $$K.E. \propto T \qquad ...(10)$$

Hence the kinetic energy of one mole of any gas is directly proportional to its absolute temperature.

This is also the Maxwell's generalisation, which states that "*Kinetic energy of translation of an ideal gas is independent of the nature of the gas, its pressure and depends only upon the temperature.*"

But $$K.E. \propto \bar{c}^2$$

$$\therefore \quad \bar{c}^2 \propto T$$

or $$\bar{c}^2 \propto \sqrt{T} \qquad ...(11)$$

Hence molecular velocity of any gas is proportional to the square root of the absolute temperature. The molecular motion is, therefore, often known as *thermal motion of the molecules.*

At absolute zero, the thermal motion ceases completely and kinetic energy of the gas would become also zero.

Deviation of Gas Laws from the Kinetic Equation

(a) Boyle's Law: *This law states, "Temperature remaining constant, the volume of a give mass of a gas is inversely proportional to the pressure."* mathematically it can be expressed as

$$V \propto \frac{1}{P}$$

$$PV = \text{constant}$$

This law can be derived from the kinetic equation

$$PV = \frac{1}{3}mN\bar{c}^2$$

This can be written as

$$PV = \frac{2}{3} \times \frac{1}{2}mN\bar{c}^2 \qquad ...(i)$$

But $\frac{1}{2}mN\bar{c}^2$ = average kinetic energy of gas which is proportional to the absolute temperature.

i.e. $$\frac{1}{2}mN\bar{c}^2 \propto T$$

or $$\frac{1}{2}mN\bar{c}^2 = KT$$

Substituting in (i) above

$$PV = \frac{2}{3}.KT$$

In this equation $\frac{2}{3}KT$ would be a constant for a give mass of gas at a constant temperature.

$\therefore$ $$PV = \text{constant}$$

or $V \propto \frac{1}{P}$. This is Boyle's law.

(b) Charle's Law: *According to this law, "The volume of a give mass of a gas is directly proportional to the absolute temperature at a constant pressure."*

We have already derived

$$PV = \frac{2}{3}kT$$

or $$V = \frac{2}{3} \times \frac{k}{P}.T$$

or $\frac{V}{T} = \frac{2k}{3P}$ Thus for a give mass of a gas if pressure P, remains constant, the expression $\frac{2k}{3P}$ is constant.

$\therefore$ At constant pressure

$$\frac{V}{T} = K$$

or $$V = K.T.$$

Hence $V \propto T$. This is Charle's law.

(c) Avogadro's Hypothesis: *This law states, "Equal volumes of all gases under similar conditions of temperature ad pressure contain equal number of molecules."*

For any two gases, the kinetic equation can be put as follows:

(i) $$P_1V_1 = \frac{1}{3} m_1 N_1 \bar{c}_1^2$$

(ii) $$P_2V_2 = \frac{1}{3} m_1 N_1 \bar{c}_1^2$$

On rewriting:

(i) $$P_1V_1 = \frac{2}{3} \times \frac{1}{2} m_1 N_1 \bar{c}_1^2$$

(ii) $$P_2V_2 = \frac{2}{3} \times \frac{1}{2} m_1 N_1 \bar{c}_2^2$$

Since $P_1 = P_2$ and $V_1 = V_2$

$$\therefore \quad \frac{1}{2} m_1 N_1 \bar{c}_1^2 = \frac{1}{2} m_1 N_1 \bar{c}_2^2 \qquad ...(a)$$

Now if the temperature is also constant, the average kinetic energy per molecule must be the same.

or $$\frac{1}{2} m_1 \bar{c}_1^2 = \frac{1}{2} m_2 \bar{c}_2^2 \qquad ...(b)$$

On dividing (a) by (b), we have

$$N_1 = N_2$$

This is Avogadro's Law

(d) The Ideal Gas Equation: On Combining Boyle's law, Charles law ad Avogadro's law we find that the volume of a gas depends on the pressure, temperature and number of moles, as follows:

$V \propto 1/P$ (at constant T and n) (Boyle's law)

$V \propto T$ (at constant P and n) (Charles' law)

$V \propto n$ (at constant T and P) (Avogadro's law)

Thus, V should be proportional to the product of these three terms, *i.e.*,

$$V \propto nT/P$$

or
$$V = R(nT/P)$$

or
$$PV = RT \qquad ...(i)$$

where R, the proportionality constant, is called the gas constant. Eq. (i) is called the ideal gas equation.

The Gas Constant: As one mole of a ideal gas occupies a volume of 22.414 dm^3 at N.T.P. (0°C ad 1 atmp pressure), we find from Eq. (i) the gas constant

$$R = \frac{PV}{nT} = \frac{(1\ atm)(22.414\ dm^3)}{(1\ mol)(273.15\ K)}$$

$$= 0.08206\ dm^3\ atm\ K^{-1}\ mol^{-1}$$

In SI units, 1 atom = $1.01325 \times 10^5\ N\ m^{-2}$, so that

$$R = \frac{PV}{nT}$$

$$= \frac{(1.03125 \times 10^5\ N\ m^{-2})(22.414 \times 10^{-3}\ m^3)}{(1\ mol)(273.15\ K)}$$

$$= 8.314\ N\ m\ K^{-1}\ mol^{-1}$$

$$= 8.314\ J\ K^{-1}\ mol^{-1} \qquad (\because\ J = Nm)$$

Note: 1 dm^3 = $1 \times 10^{-3}\ m^3$ = 1000 cm^3 = 1 litre.

(e) Dalton's Law of partial pressure: This law states, "*Total pressure P of a number of gases i a enclosed space is equal to the sum of the individual (partial) pressures of each gas, provided the gases do not react with one another.*" Thus

$$P = p_1 = p_2 + p_3 + p_4 + ... + p_N$$

Gas pressure is the force per unit area exerted by the molecules of the gas against the walls of the container.

Suppose N_1 molecules, each of mass m_1 of a gas A, are contained in a vessel of volume V.

The according to the kinetic theory equation the pressure is given by

$$p_a = \frac{m_1 N_1 \bar{c}_1^2}{3V} \qquad ...(i)$$

where $\bar{c}_1$ is the root mean square velocity of the molecules of the gas A.

Now, suppose N_2 molecules, each of mass m_2 of another gas B, are contained i the same vessel at the same temperature and there is no other gas present at that time. The pressure of the gas will b given by

$$p_b = \frac{m_2 N_2 \bar{c}_2^2}{3V} \qquad ...(ii)$$

where $\bar{c}_2$ is root mean square velocity of the molecules of the gas B.

If both the gases are present in the same vessel, the total pressure P will be give by

$$P = \frac{m_1 N_1 \bar{c}_1^2}{3V} + \frac{m_2 N_2 \bar{c}_2^2}{3V} = p_a + p_b$$

Similarly if three, four or ore gases were present, the total pressure P will be given by

$$P = p_a + p_b + p_c + p_d \cdots$$

This is Dalton's law of partial pressures.

(f) Gaham's Law of Gaseous Diffusion: The law may be stated "*Under similar conditions of temperature and pressure, the rates of diffusion of gases are inversely proportional to the square roots of their densities.*"

Mathematically,

$$\frac{r_1}{r_2} = \sqrt{\frac{D_2}{D_1}}$$

According to the kinetic equation

$$PV = \frac{1}{3} m N \bar{c}^2$$

or

$$\bar{c} = \sqrt{\frac{3PV}{mN}}$$

$$= \sqrt{\frac{3PV}{M}} \quad \text{[mN = molecular weight of the gas.]}$$

$$= \sqrt{\frac{3P}{D}} \qquad [\because \quad D = \frac{M}{V}]$$

$$\bar{c} = \sqrt{\frac{1}{D}} \text{ if P is kept constant.}$$

$\bar{c}$ is the mean velocity of the molecules and rate of diffusion of a gas will be dependent upon it at a particular temperature.

$$r \propto \bar{c}$$

$$r \propto \sqrt{\frac{1}{D}} \text{ if P is kept constant.}$$

$\bar{c}$ is the mean velocity of the molecules and rate of diffusion of a gas will be dependent upon it at a particular temperature.

$$r \propto \bar{c}$$

$$r \propto \sqrt{\frac{1}{D}}$$

This is Graham's Law.

CALCULATIOΝ OF MOLECULAR VELOCITY

In calculating various properties of the gases, one makes use of the following three velocities.

(i) Root mean Square velocity (RMSV): *It may be defined as the square root of the mean of the velocities of the molecules.* If v_1 v_2 ... etc. are the velocities possessed by different molecules, then

$$\sqrt{\bar{c}^2} = \sqrt{\frac{v_1^2 + v_2^2 + ... + v_N^2}{N}}$$

The expression to calculate this velocity can be obtained from the kinetic gas equation, according to which

$$PV = \frac{1}{3} mN\bar{c}^2$$

For 1 mole of the gas

$$PV = RT \text{ and } N = NA \text{ thus}$$

$$\sqrt{\bar{c}^2} = \sqrt{\frac{3PV}{mNa}} = \sqrt{\frac{3RT}{mNa}} = \sqrt{\frac{3RT}{M}} = \sqrt{\frac{3kT}{m}} = \sqrt{\frac{3P}{d}}$$

where K is Boltzmann constant k = R/Na

(2) Average Velocity (c): *It is defined as the average of the various velocities possessed by the molecules, i.e., as*

$$\bar{c} = \frac{v_1 + v_2 + \ldots + v_N}{N}$$

It can be calculated by using the expression

$$\bar{c} = \sqrt{\frac{8RT}{\pi M}}$$

$$\bar{c} = \sqrt{\frac{8RT}{\pi M}} = \sqrt{\frac{8kT}{\pi m}}$$

(3) Most Probably Velocity (MPV): It may be defined *as the velocity corresponding to the maximum fraction of the molecules.* It is given by the following expression

$$MPV = \sqrt{\frac{2RT}{M}} = \sqrt{\frac{2PV}{M}} = \sqrt{\frac{2P}{d}}$$

The relations between these velocities can be easily worked out

$$\frac{MPV}{\bar{c}^2} = \frac{\sqrt{2RT/M}}{\sqrt{8RT/\pi M}} = \sqrt{\frac{22}{28}} = \frac{1}{1.128}$$

$$\frac{MPV}{RMSV} = \frac{\sqrt{2RT/M}}{\sqrt{3RT/M}} = \sqrt{\frac{2}{3}} = \frac{1}{1.224}$$

Thus MPV : $\bar{c}$: RMSV : : 1 : 1 . 128 : 1.224.

i.e. RMSV > c > MPV. The relation between average and RMSV will be

$$\text{Average} = \left(\frac{1.128}{1.224}\right) MPV = 0.9213\ MPV.$$

Expansivity and Compressibility

When fluids (gases and liquids) are heated, they expand but the expansion of gases is much more than that of liquids. Similarly, they can be compressed. The variation of volume V with temperature T, keeping pressure P constant is termed as coefficient of thermal expansion or the coefficient of isobaric expansion or simple, expansivity, α of the fluid.

Thus,

$$\alpha = \frac{1}{V}\left(\frac{\partial V}{\partial T}\right)_P \qquad ...(1)$$

Similarly, the variation of V with P, keeping T constant, is termed *as the coefficient of isothermal compressibility* or, simply, compressibility, β of the fluid, thus,

$$\alpha = -\frac{1}{V}\left(\frac{\partial V}{\partial T}\right)_T \qquad ...(2)$$

Generally, as the volume of a gas gets decreased as pressure increases it means that the quantity $(\partial V/\partial P)_T$ is negative. Hence, a minus sign is included in the definition so as to make β positive. It is to be noted that α has the dimensions of T^{-1} and β those of P^{-1}.

Let us derive the value of α and β for an ideal gas.

For n moles of an ideal gas, we have

$$PV = nRT \qquad ...(i)$$

or

$$V = nRT/P \qquad ...(ii)$$

On differentiating Eq. (ii) with respect to T at constant P, we get,

$$\left(\frac{\partial V}{\partial T}\right)_P = \frac{nR}{P} = \frac{PV}{TP} = \frac{V}{T}$$

$$\therefore \quad \alpha = \frac{1}{V}\left(\frac{\partial V}{\partial T}\right)_P = \left(\frac{1}{V}\right)\left(\frac{V}{T}\right) = \frac{1}{T}$$

Again, differentiating Eq. (ii) with respect to P at constant T. We get,

$$\left(\frac{\partial V}{\partial P}\right)_T = \frac{nRT}{P^2} = \left(\frac{nRT}{P}\right)\left(-\frac{1}{P}\right) = -\frac{V}{P}$$

$$\therefore \quad \beta = -\frac{1}{V}\left(\frac{\partial V}{\partial P}\right)_T = \left(-\frac{1}{V}\right)\left(-\frac{V}{P}\right) = \frac{1}{P}.$$

CRITICAL CONSTANTS

(i) *Critical temperature of a gas (T_c)* may be defined as the minimum temperature which must be reached before it can be liquefied by the application of pressure. A gas cannot be liquefied by pressure above its critical temperature, no matter how high the applied pressure may be. Each gas is having definite critical temperature, *e.g.*, critical temperature of CO_2 is 31.1MC.

(ii) *Critical pressure of a gas* (P_c) may be defined as the minimum pressure necessary for the liquefaction of a gas at critical temperature.

(iii) *Critical volume of a gas* (V_c) may be defined as the volume occupied b 1 mole of the gas or liquid under critical pressure and under critical temperature.

CONDITIONS FOR LIQUEFACTION OF GASES

From the Andrew's experiment very useful conditions for the liquefaction of a gas may be stated as :

(i) The temperature must be at or below the critical temperature.

(ii) A suitable pressure must be applied equal to the critical pressure if the temperature is at the critical temperature but becoming less if the temperature is reduced below this value.

Van der Waals Equation as an Improvement Over Simple Laws

Van der Waals equation is an improvement over simple laws due to following reasons:

(i) *Van der Waals equation explains the devication from Boyle's law:* Regnault and Amagat studied the behaviour of cases and observed that they do not obey simple gas laws under different conditions of temperature and pressure, more particularly at low temperature and high pressure. They also plotted graphs between the product of pressure and volume ad pressure and found that PV product was not constant as desired by Bole's law. These deviations from ideal behaviour were explained by van der Waals equations.

(ii) *It also explains critical phenomenon.* This equation satisfactorily explains critical phenomenon, observed in the case of gases.

The theoretical isotherm (*i.e.*, the curves between V and P at constant T) obtained from the values of V and the experimental values of 'a' and 'b' from van der Waals equation were found to be quite similar. The theoretical isotherms obtained from simple gas laws are quite different from the experimental isotherms.

DEVIATION FROM IDEAL GAS BEHAVIOUR

If we study the actual behaviour of the gases, it can be see that no gas *is ideal or perfect* in this sense although some gases tend to approach perfection as the temperature rises above their boiling points.

Since we know that no gas is ideal or perfect, the equation PV = nRT is obeyed as an approximation by real gas at low pressure while at high pressure and low temperature there occurs much deviation from this ideal behaviour.

Regnault and Amagat attempted to study the effect of change of P on PV of real gases like H_2, He, N_2O_2 and CO_2 under different conditions of temperatures and demonstrated that they did not obey Boyle's law, more particularly at high pressure and low temperatures.

From the diagram the following conclusions can be drawn:

(i) *At low Pressure:* Left hand portions of the curves show the devications at low pressures. With the rise of pressure, PV increases for hydrogen and helium whereas it decreases for 23 and CO2. Same is the case with other gases. *Thus at low pressure PV for all gases excepting H_2 and He, is having a lower value than expected of an ideal gas.*

(ii) *At high pressure:* Right-had portions of the curves show the deviations at high pressures. For H_2 and He, PV gets increased regularly with the rise of pressure but for N_2 and CO_2 it first decreases, passes through a minimum where it remains constant for a while and the rises. At very high pressure PV for all gases is having a higher value than that expected of an ideal gas.

Thus at high pressures PV for all gases including H_2 and He is having higher value than expected for an ideal gas.

The above results can be interpreted as follows:

(a) At moderate pressures, there occur a negative or under perfect deviation. Hydrogen and helium do not appear to show under perfect behaviour at ordinary temperature. However, at a much lower temperature hydrogen and helium also give a curve of the same type as the other gases.

(b) All the curves approach the ideal value as the pressure approaches zero, *i.e.*, at low pressure all gases behave as ideal gases.

(c) At very high pressures, there occurs a positive or over perfect devication.

(d) At very high pressures gases do not eve approximately obey gas laws.

(iii) *Effect of temperature:* There are two sets of curves, one at 170186 and the other at 100°C. It is evident from these curves that the deviations are greater at 17°C while in case of CO_2 a sharp break occurs as the liquefaction point gets reached.

In general if the temperature gets changed, the shape of the curve would remain the same but the *deviations would become more pronounced as the temperature falls.*

The extent of under perfect behaviour is more or less dependent on how near the gas is to the temperature at which it will liquefy. This explains why CO_2 exhibits much greater deviation than N_2 while H_2 (which liquefies at a much lower temperature) exhibits over perfect behaviour.

It means that deviations decrease with increase in temperature. It may be emphasized that Boyle's law is obeyed only at very low pressure and moderately high temperature.

For every gas there occurs a characteristic temperature when PV-P curve runs considerably parallel to the pressure axis from zero to moderate values of pressure. It implies that *at this temperature the value of PV remains constant for an appreciable range of pressure* signifying that at this temperature Boyle's law gets obeyed for a considerable range of pressure. This temperature is termed as the Boyle's temperature.

Explanation of Deviation

So as to explain deviations from ideal behaviour, it becomes essential to modify the kinetic theory of gases. The following two postulates of the kinetic theory do not hold good under all conditions. We will make an attempt to examine them more critically.

Postulate No. 1: The volume occupied by the gas molecules themselves is negligibly small as compared to the total volume occupied by the gas. One can justify this postulate only under ordinary conditions of temperature and pressure.

By calculations it can be see that in some of the common gases, the volume occupied by the molecules themselves, under ordinary conditions, is early 0.014 percent of the total volume of the gas. This is a negligible fraction. But, if the pressure becomes too high (say, 100 atmosphere or more), the total volume of the gas will decrease appreciably while the volume of the molecules will remain almost the same because the molecules are incompressible. *Therefore under conditions of high pressure, the volume occupied by the gas molecules cannot be neglected in comparison with the total volume of the gas.*

The same thing is expected to happen when the temperature is decreased to a large extent. The total volume of the gas gets decreased considerably, no doubt, but the volume occupied by the molecules themselves remains practically the same.

In this case, too, *the volume occupied by the molecules would be no longer be negligible. Thus the postulate No. 1. is not valid at high pressures ad low temperatures.*

Postulate No. 2 : The forces of attraction between gas molecules are negligible. This assumption is expected to be valid at low pressures or at high temperatures because under these conditions the molecules li far apart from once another.

But at high pressures or at low temperatures, the volume becomes small and molecules lie closer to one another. The intermolecular forces of attraction, therefore, become appreciable and cannot be ignored.

Hence, *the postulate No. 2, is not valid under conditions of high pressure and low temperature.*

It becomes necessary, therefore, to apply suitable corrections to the ideal gas equation so as *to make it applicable to real gases.*

Explanation of behaviour of real gases of the basis of van der Waals equation

$$\left(P + \frac{a}{V^2}\right)(V - b) = RT \quad \text{...van der Waals equation of state}$$

or $$PV + \frac{a}{V} - Pb + \frac{ab}{V^2} = RT$$

On Neglecting $\frac{ab}{V^2}$ a product of two very small quantities, we obtain

$$PV - + \frac{a}{V} = RT = P'V'$$

where P′ and V′ refer to the ideal pressure and volume of the gas respectively.

(i) At low pressure: Small addition of pressure due to mutual attraction plays an important part while b is negligible as compared to large total volume V. The factor a/V, therefore overweights Pb *i.e.*,

$$\frac{a}{V} > Pb$$

$$\frac{a}{V} - Pb = Z, \text{ a positive quantity.}$$

From eq. (1), we have

$$PV + Z = RT = P'V'$$

$$PV = RT - Z = P'V' - Z$$

i.e., the observed PV is less than P′V′.

The observed product PV at low pressure would be less than P′V′, the product of pressure and volume if the gases were ideal.

At low pressures, therefore PV decreases with increase in pressure in case of N_2, O_2 and CO_2, etc.

(ii) At high pressure: If pressure is high, the volume is small and small volume occupied by molecules themselves cannot be neglected. Whereas in comparison with the high pressure, small addition of pressure due to mutual attraction is negligible. Hence Pb overweights a/V, *i.e.*,

$$Pb > \frac{a}{V}.$$

i.e., $\frac{a}{V} - Pb = -y$, a negative quantity

$$\therefore \quad PV + (-y) = P'V'$$

$$PV - y = P'V'$$

$$PV = P'V' + y$$

i.e., the observed PV is greater than the value expected from an ideal gas.

This explains the fact that PVl, after reaching a minimum, increases with further increase in pressure.

(iii) At ordinary (medium) pressure: It is seen that at ordinary temperature the factor $\frac{a}{V^2}$ becomes predominant at low pressures whereas the factor Pb predominates at high pressures. Therefore, these two factors will balance each other at some intermediate range of pressures in which the gas will show the ideal behaviour.

Hence at medium pressure, $Pb = \frac{a}{V}$

The equation (1) is reduced to

$$PV = RT = P'V'$$

and the gas behaves like an ideal gas.

(iv) At high temperature: At high temperatures, V is very large and hence both Pb and $\frac{a}{V}$ will be negligibly small, so that we have

$$PV = RT = P'V'$$

i.e., at high temperatures the behaviour of a real gas nearly agrees with the expected for ideal gas.

This explains why the deviations become less at high temperature.

(v) At low temperature: Both P and V are small, hence both pressure (attraction) and volume corrections are appreciable, with the result that the deviations are more pronounced.

(vi) Exceptional behaviour of hydrogen and helium: The molecules of these gases are of very small mass and consequently the force of attraction between the molecules is always negligible.

Hence neglecting the term $\frac{a}{V}$ equation (1) gets reduced to

$$PV - Pb = RT = P'V'$$

$$PV = P'V' + Pb$$

i.e., the product PV would become greater than P'V' event at low pressures.

EXPERIMENTAL DETERMINATION OF CRITICAL CONSTANTS

Determination of Critical Temperature (T_c) and Critical Pressure (P_c)

Andrews determine critical temperature and critical pressure of a gas from its isotherms at different temperatures. This is a simple method which is based on the principle that *at critical temperature the surface of separation between liquid and vapour phases disappears.* This method is used when the substance is in liquid state at ordinary temperature. The apparatus consists of a U-shaped hard glass tube having one limb smaller than the other. The tube is filled partially with mercury enclosing a small quantity of dry air in the longer limb. The upper portion of the longer tube is graduated.

A small quantity of liquid and its vapour is kept over the mercury in the shorter limb and the mouth is sealed. The portion of to he shorter tube having liquid vapour, is then enclosed in a jacket, the temperature of which may be changed slowly and gradually when desired by circulating a suitable liquid from a thermostat.

The temperature of jacket is gradually increased. As the liquid is being heated, its surface tension is slowly decreasing and the boundary (or meniscus) would become progressively flatter and flatter. When the critical temperature is reached, the boundary or meniscus between the liquid and the vapour flickers and suddenly would disappear. The temperature at which this takes place is noted. The tube is then cooled a bit slowly and temperature is again noted when the gas becomes turbid for a moment and a small amount of liquid quickly settles at the bottom with the subsequent reappearance of the meniscus. The mean of the two temperature radians is regarded to be the critical temperature (T_c). At the critical temperature the reading of the corresponding pressure from the manometer is also noted. The mean of the two pressure reading corresponding to the two temperatures would five the critical pressure (P_c).

Determination of Critical Volume (V_c)

It is not possible to determine the critical volume (V_c) of the gas experimentally because slight changes in pressure and temperature of the gas give rise to large changes in volume. So it becomes difficult to determine the V_c accurately by experiment.

However, the critical volume (V_c) can be determined indirectly by using the well known rule of rectilinear diameter due to Cailletet Mathias known as Cailletet and mathias rule which states that *the mean value of densities of any substance in the liquid state and its saturated vapour at the same temperature is a linear function of temperature i.e., the mean of the densities of liquid and its saturated vapour would be lying on a straight line called the rectilinear diameter.*

Thus if the density of a liquid is plotted against temperature, a curve of the type BC. If the density of the saturated vapour of the liquid is plotted against temperature, curve AC is obtained. The two curves meet at a point C which must correspond to the density at critical temperature. At point C, the liquid and its vapour become co-existent and identical and this point must give the critical density. In practice some difficulty arises in getting the upper portions of the two curves as the conditions existing there are uncertain.

Hence, the upper portions of the curves would be almost approximate. Consequently the densities of the liquid and its saturated vapours would be determined at a various temperatures and plotted against T up to a few degrees below T_c (critical temperature). The curve is then made to complete.

The mean densities are then plotted against the various temperatures (T) when a rectilinear (straight) line DC is obtained. The rectilinear diameter is then made to extend to meet the curve at C. The point C, therefore, corresponds to T_c (critical temperature) and critical density (D_c) of the substance. The value V_c is then obtained as follows:

$$\text{Critical volume } (V_c) = \frac{\text{Mol. wt. of sustance}}{\text{Critical density}} = \frac{M}{D_c}$$

Critical constants of some common gases are give in Table 3.

Table 3 : Critical constants of gases.

Gas	*Critical Temperature, °K*	*Critical Pressure, atm*	*Critical Volume, ml.*
Helium, He	5.2	2.26	60.68
Hydrogen, H_2	33.2	12.28	64.51
Nitrogen, N_2	126.0	33.5	90.03
Carbon monoxide, CO	134.4	34.6	90.03
Argon, Ar	150.7	48.2	78.43
Oxygen, O_2	154.3	49.7	74.42
Methane, CH_4	190.2	45.6	98.76
Carbon dioxide, CO_2	304.3	72.2	95.65
Nitrous oxide, N_2O	306.9	71.9	96.91
Hydrochloric acid, HCl	324.5	81.6	89.90
Hydrogen sulphide, H_2S	373.5	89.0	126.86
Ammonia, NH_3	405.5	111.5	72.03
Chlorine, Cl_2	417.1	76.1	123.91
Sulphur dioxide, SO_2	430.3	77.6	124.75

The P-V Isotherms of Carbon Dioxide—Critical Phenomenon

The curves representing the variation of volume and pressure at constant temperature are termed as Isotherms. (Greek : isos = 'equal'). In the case of an ideal gas the isotherms are almost rectangular hyperbolas. In 1869 T. Andrews attempted to study the effect of pressure on volume of CO_2 at different temperatures. In this experiment, some CO_2 gas was enclosed in a capillary tube and the volume occupied by the gas was measured at different pressures upto about 400 atmospheres.

His apparatus consisted of a sort of hydraulic press. Two similar capillaries were taken. One was filled partially with carbon dioxide and a pellet of mercury was introduced in the tube to act as a seal. In the other capillary, air was filled and sealed by mercury. These capillaries were then fixed into metal compression chambers by liquid tight joint. The chambers were interconnected ad carried screw plugers at the bottom to apply pressure.

The upper portions of the tubes were surrounded by water bath to keep the temperature constant. If the volume readings were plotted against the pressure a set of curves of the type were obtained. Each curve would correspond to one particular temperature and is, therefore, called isotherm. The curves should be rectangular hyperbolas like that expected for an ideal gas. It is quite evident that CO_2 does not yield rectangular hyperbolas although with increasing temperature the curves approximate to one. We will now consider the various isotherms one by one.

(a) *At 13.1°C:* From the isotherm it can be seen that carbon dioxide is entirely gaseous at low pressures (*e.g.*, along AB) ad the volume decreases with increasing pressure in the normal fashion (*i.e.*, AB is roughly hyperboloid). The point A would correspond to gaseous CO_2 occupying a certain volume under a certain pressure. At low pressures, if the pressure gets increased, the volume would decrease as shown by the curve AB.

At B liquefaction starts; vapour gets converted into liquid of very much greater density and the volume gets decreased suddenly as the volume occupied by a liquid would be much less than that of gas so that BC would be almost parallel to the volume axis-indicating the sudden decrease in volume occurring at the same pressure (49.8 atmospheres) as more and more of the gas changes into liquid.

Liquefaction gets completed at C and CD reveals the small effect of pressure on a liquid. The steep line CD which is almost vertical represents the compressibility of a liquid and reveals that the liquids are much less compressible, *i.e.*, increase in pressure will produce a very small decrease in volume.

In order to summaries, along curve AB there is only vapour, along CD only liquid and along BC liquid and vapour co-exist. As BC is parallel to the volume axis, there can be only one pressure at which liquid and vapour co-exist. This pressure is termed as the *vapour pressure* of the gas.

(b) The next isotherm EFGH at 21.1°C is also similar with the difference that the horizontal portion where liquefaction occurs is slightly short (*i.e.*, the range of volumes over which gas and liquid can co-exist is smaller) and the vapour pressure higher (*i.e.*, the pressure at which liquefaction commences is higher).

(c) As the temperature gets raised, this horizontal portion would become less and less till at 31.1°C it reduces almost to a point X. Above this temperature there occurs no trace of horizontal portion, *i.e.*, beyond 31.1°C there occurs no indication of liquefaction at all.

Anderws applied a pressure of 300 to 400 atmospheres but could to liquefy CO_2 if the temperature was beyond 31.1°C. Below 31.1°C a pressure of 75 atmospheres was quite sufficient to liquefy carbon dioxide.

Point X refers to the critical point for CO_2. The distinction between liquid and vapour state disappears at this point and CO_2 exists in a state called the critical state. The isotherm passing through critical point is termed as the critical isotherm.

From the experiment Andrews was able to conclude that if the temperature of CO_2 is above 31.1°C it cannot be liquefied no matter how high the pressure may be. This maximum temperature (31.1°C) at which liquefaction can be brought about is termed as critical temperature.

The pressure required to liquefy a gas at the critical temperature is termed as the critical pressure.

If the temperature is below the critical temperature, the gas will liquefy at a somewhat lower pressure. All other gases have been found to behave similarly. The critical phenomenon observed b Andrews for CO_2 can be observed for any gas.

COMPRESSIBILITY FACTOR

It is known that the equation of state, PV = nRT, derived from the postulates of the kinetic theory, *is valid for an ideal gas only*. Real gases tend to obey this equation only approximately ad that too under conditions of low pressure and high temperature.

The higher the pressure and the lower the temperature, the greater would be the deviations from the ideal behaviour. In general, the most easily liquefiable and highly soluble gases exhibit larger deviations. Hence gases like carbon dioxide, sulphur dioxide and ammonia exhibit much larger deviations than hydrogen, oxygen, nitrogen, etc.

It is best to represent the deviations from ideal behaviour i terms of the compressibility factor, Z, which is defined as

$$Z = \frac{PV}{(PV)_{ideal}} = \frac{PV}{nRT} \quad ...(1)$$

For an ideal gas, Z = 1 under all conditions of temperature and pressure. The deviation of Z from unity is, therefore a measure of the imperfection of the gas under consideration.

The graphs plotted for the compressibility factors determined for a number of gases over a range of pressures at a constant temperature (0°C). At extremely low pressures, the value of Z for all the gases are close to unity which implies that the gases behave almost ideally. At very high pressures, all the gases have Z more than unity indicating that the gases are less compressible than an ideal gas. This is attributed to the fact that at high pressures, the molecular repulsive forces are dominant.

Moderately low pressures, carbon monoxide, methane and ammonia are more compressible than an ideal gas, *i.e.* $Z < 1$. This is attributed to the fact that at low pressures, the long range attractive forces are dominate and favour compression. The compressibility factor Z goes on decreasing with increase in pressure, passes through a minimum at a certain stage and then starts increasing with increase in pressure. The gases would now becomes less compressible than ideal gas, *i.e.*, $Z > 1$. It can be seen from the figure that although, carbon monoxide and methane exhibit marked devications from ideal behaviour only at high pressures, yet ammonia exhibits large deviation even at low pressure.

It can be seen that hydrogen and helium at 0°C are less compressible *at all pressures* than the ideal gas, *i.e.*, $Z > 1$. However, if the temperature is sufficiently low (*e.g.*, below – 165°C for hydrogen and below – 240°C for helium) these gases would also give the same type of Z – P plots as are shown by ammonia, carbon monoxide and methane at 0°C.

On the other hand, if the temperature becomes sufficiently high, the Z-P plots of ammonia, carbon monoxide and methane will be almost similar to those of hydrogen and helium at 0°C *i.e.*, the value of Z will increase continuously with increasing pressure.

PRINCIPLE OF CONTINUITY OF STATES

Although, the transition from liquid to vapour and *vice versa* generally takes place suddenly and the two states can be sharply distinguished from each other as can be see from AB and CD parts of 13.1°C isotherm for carbon dioxide, yet where occurs a possibility of passing from one state of aggregation to the other without involving sudden change.

Hence for further theoretical considerations it becomes essential to show that the liquid stated does not refer to a sharp and discontinuous transition from gaseous state but rather a continuation of the gaseous phase into the region of very strong intermolecular attractions where the volume becomes very small. The idea of a contiguity from liquid to gaseous state will become evident from the following consideration: The dome-shaped curve depicted b the dashed line connects the end points of the horizontal portions of the isotherms.

As a horizontal portion represents two phases in equilibrium, all points inside the dome correspond to equilibrium between liquid and vapour. For such points it always becomes possible to draw any sharp distinction between liquid and vapour, because there is a definite boundary surface between the two phases.

For points lying outside the dome only there is only one phase (liquid or vapour) ad here it is not possible to draw any sharp distinction between liquid and vapour. This fact is termed as the *principle of contiguity of states*.

Suppose, that carbon dioxide gas under atmospheric pressure is made to heat a temperature above the critical temperature, as to 48°C and that is then subjected to a pressure greater than critical pressure, as, to 100 atmospheres, the system will change volume as shown by the isotherm 48°C.

When the temperature is lowered to 13.1°C while maintaining the pressure at 100 atmospheres, a point on the curve CD will be reached a curve which applies to liquid carbon dioxide. At no point in the process there is any abrupt change of state; the change from gas to liquid will be contiguous. Similarly, it is possible to start with liquid carbon dioxide under the conditions represented, say by the point C.

If we now raise the pressure o the liquid to 10 atmospheres and the temperature to 48°C the condition of the system will be represented by a point on that isotherm; and on reducing the pressure to that of the atmosphere ad then allowing the temperature to fall to 13.1°C, we shall be having carbon dioxide as a gas; and at no time during the whole process any discontinuity of properties will be observed, or an separation into liquid and vapour. The passage from the liquid to the gaseous state would be continuous.

An important conclusion derived from the principle of continuity of states is that one may think of a liquid near the critical point as just a highly compressed gas. If a equation of state can be found which accurately describes the gas close to its critical temperature and pressure that equation should also be able to describe the liquid.

Van der Waals Equations and Critical State—Relations between Van der Waals Constants and Critical Constants

A very good confirmation of the general character of van der Waals equation may be provided b its application to critical phenomenon.

The van der Waals equation may be put as follows:

$$\left(P + \frac{a}{V^2}\right)(V - b) = RT \text{ (putting } = 1) \quad ...(1)$$

$$PV + \frac{a}{v} - Pb - \frac{ab}{V^2} = RT$$

On multiplying throughout by V^2, we obtain

$$PV^3 + aV - PbV^2 - ab = RTV^2$$

or $$PV^3 + aV - PbV^2 - ab - RTV^2 = 0$$

On arranging in descending powers of V, we get

$$PV^3 - PBV^2 - RTV^2 + aV - ab = 0$$

On dividing b P, we get

$$V^3 - bV^2 - \frac{RTV^2}{P} + \frac{aV}{P} - \frac{ab}{P} = 0$$

$$V^3 - V^2 - \left(b + \frac{RT}{P}\right) + \frac{aV}{P} - \frac{ab}{P} = 0 \qquad ...(2)$$

This is a third degree equation in terms of V (cubic equation in V) and below a certain value of temperature T the equation can be having one of three real roots. With the rise of temperature these roots would approach one another and become identical ultimately.

This can be readily understood if a theoretical P – V graph is plotted from the van der Waals equation. The three real unequal roots correspond to point B, F and C. The points B and C represent saturated vapour and liquid respectively, but F is having no physical significance.

As the temperature gets raised the S-shaped dotted portion of the curve gets reduced to a point. The values for the volume of liquid and the gas approach each other until at this volume of liquid and the gas approach each other until at this point, the three values of V become identical. This point would correspond to the critical temperature T_c and the volume V_c.

At the critical point, the three roots of the van der Waals equation would become identical and equal to V_c. But this refers to the value of the three roots of the van der Waals equation at the critical point.

$$(V - V_c)^3 = 0 \qquad ..(3)$$

or

$$V^3 - 3V_cV^2 + 3V_c^2V - V_c^3 = 0 \qquad ...(4)$$

On substituting these values in eq. (2), we get,

$$V^3 - V^2\left(b + \frac{RT_c}{P_c}\right) + \frac{aV}{P_c} - \frac{ab}{P_c} = 0 \qquad ...(5)$$

As both eqs. (4) and (5) are identical then the coefficients of the equal powers of V i the two equations may be equated:

$$3V_c = b + \frac{RT_c}{P_c} \qquad ...(6)$$

$$3V_c^2 = \frac{a}{P_c} \qquad ...(7)$$

$$V_c^3 = \frac{ab}{P_c} \qquad ...(8)$$

On dividing Eq. (8) by Eq. (7), we get,

$$\frac{V_c^3}{3V_c^2} = \frac{ab}{P_c} \times \frac{P_c}{a}$$

or $$\frac{V_c}{3} = b$$

$$V_c = 3b \quad ...(9)$$

Substituting the value of V_c from Eq. (9) into Eq. (7), we obtain

$$\frac{a}{P_c} = 3 \times (3b)^2$$

or $$\frac{a}{P_c} = 27b^2$$

or $$P_c = \frac{a}{27b^2} \quad ...(10)$$

Substituting the values of V_c and P_c in equation (6), we get

$$3 \times 3b = b + \frac{RT_c}{a/27b^2}$$

$$9b = b + \frac{RT_c}{a/27b^2}$$

or $$8b = RT_c \times \frac{27b^2}{a}$$

or $$T_c = 8b \times \frac{a}{27b^2} \times \frac{1}{R}$$

$$T_c = \frac{8a}{27bR} \quad ...(11)$$

Derivation of $P_cV_c = \frac{3}{8} RT_c$ from van der Waals Equation and calculation of van der Waals constants 'a' and 'b' in terms of T_c and P_c.

From van der Waals equation, we get

$$V^3 - V^2\left(b + \frac{RT_c}{P_c}\right) + V\frac{a}{P_c} - \frac{ab}{P_c} = 0 \quad ...(12)$$

and $$V^3 - 3V_cV^2 + 3V_c^2V - V_c^3 = 0 \quad ...(13)$$

On equating the coefficients of (12), and (13), we get

$$3V_c = \frac{RT_c}{P_c} + b \quad ...(14)$$

$$3V_c^2 = \frac{a}{P_c} \qquad ...(15)$$

$$V_c^3 = \frac{ab}{P_c} \qquad ...(16)$$

On dividing Eq. (16) by (15), we get

$$V_c = 3b \quad \text{or} \quad b = \frac{1}{3}V_c \qquad ...(17)$$

On substituting the value of b in (14), we obtain

$$3V_c = \frac{RT_c}{P_c} + \frac{V_c}{3}$$

or
$$3V_c - \frac{V_c}{3} = \frac{RT_c}{P_c}$$

or
$$\frac{8}{3}V_c = \frac{RT_c}{P_c}$$

or
$$P_cV_c = \frac{3}{8}RT_c \qquad ...(18)$$

On rearranging Eq. (18), we get

$$V_c = \frac{3}{8}\frac{RT_c}{P_c}$$

On squaring, we obtain

$$V_c^2 = \frac{9}{64}\frac{R^2T_c^2}{P_c^2} \qquad ...(19)$$

On substituting the value of $V_c^2 = \dfrac{a}{3P_c}$ from Eq. (15) into Eq. (19), we get,

$$\frac{a}{3P_c} = \frac{9}{64}\frac{R^2T_c^2}{P_c^2}$$

$$a = \frac{27R^2T_c^2}{64P_c} \qquad ...(20)$$

On substituting the value of $V_c = 3b$ in Eq. (14), we obtain,

$$9b = \frac{RT_c}{P_c} + b$$

or $$8b = \frac{RT_c}{P_c}$$

or $$b = \frac{RT_c}{8P_c} \quad ...(21)$$

SOME SOLVED PROBLEMS

Problem 1:

Oxygen has a density of 1.429 gm. per litre at N.T.P. Calculate the R.M.S. and A.V. of oxygen molecules.

Solution:

Now $$RMSV = \sqrt{\frac{3P}{D}}$$

In this case, $P = 1$ atmosphere

$$= 76 \times 13.6 \times 981 \text{ dynes/sq. cm.}$$

$$D = 1.422 \text{ gm/litre}$$

$$= \frac{1.429}{1000} \text{ gm/c.c} = 0.001429 \text{ gm/c.c.}$$

Substituting the values in the above equation, we have

$$RMSV = \sqrt{\frac{3 \times 76 \times 13.6 \times 981}{0.001429}} = 46100 \text{ cm/sec}$$

This is the R.M.S. (root mean square mean velocity) of oxygen molecules.

Now A.V. (Average velocity)

$$= 0.9213 \times \text{R.M.S. velocity}$$

$$= 0.9213 \times 46{,}100 = 42458 \text{ cm/sec.}$$

Problem 2:

A vessel of 50 litre capacity contains 10 moles of steam under 5 atmospheres of pressure. Calculate the temperature of the steam using van der Waals equation if for water $a = 5.46$ atm (litre/mole)^2 and $b = 0.031$ litre/mole.

Solution:

We know $$\left(P + \frac{n^2a}{V^2}\right)(V - nb) = nRT$$

i.e.
$$T = \left(P + \frac{n^2a}{V^2}\right)\left(\frac{V - nb}{nR}\right)$$

Here $P = 5$ atm, $n = 10$ moles

$V = 10$ litres and $a = 5.46$ atm $\left(\frac{\text{litre}}{\text{mole}}\right)^2$

and $b = 0.03$/lite/mole.

Substituting the values,

$$T = \left(5 + \frac{(10)^2 \times 5.46}{(50)^2}\right)\left(\frac{50 - 10 \times 0.031}{10 \times 0.08}\right)$$

$$= 311.9°K$$

Problem 3:

The critical pressure and critical temperature of a gas obeying van der Waals equation have values 73 atms and 37°C. Calculate constants 'a' and 'b'. The value of the gas constant, R is 0.082 atm degree ¹.

Solution:

We have
$$T_c = \frac{8a}{27Rb} \quad \text{...(i)}$$

$$P_c = \frac{a}{27b^2} \quad \text{...(ii)}$$

Dividing (i) by (ii) we have,

$$\frac{T_c}{P_c} = \frac{8a}{27Rb} \times \frac{27b^2}{a} = \frac{8b}{R}$$

Now $T_c = 31° + 273 = 304°K$

$P_c = 0.082$ atm degree^{-1}

Hence
$$\frac{304}{73} = \frac{8b}{0.082}$$

$\therefore$ $b = 0.4269$

Substituting the value of b in (i) *i.e.*,

$$T_c = \frac{8a}{27Rb}, \text{ we get}$$

$$304 = \frac{8a}{27 \times 0.082 \times 0.04269}$$

$$a = 3.591.$$

Problem 4:

Calculate the reduced pressure, volume and temperature of CH_4 gas when one mole of gas is contained in a 5 litre flask under 5 atmospheric pressure. What will be the temperature of the as when

Critical temperature $T_c = 190.2°K$

Critical volume $V_c = 0.0988$ *litre/mole and*

Critical pressure $P_c = 45.6$ *atm.*

Solution:

We know,

Reduced pressure $\pi = P/Pc$

Here P = 5 atm $P_c = 45.6$ atm.

Substituting the values, $\pi = 5/45.6 = 0.1096$

Reduced volume $\phi = V/V_c$

Here V = 5 litres and $V_c = 0.0988$ litre/mole

substituting the values $\phi = \dfrac{5}{0.0988} = 50.607$

According to the law of corresponding states,

$$\left(\pi + \frac{3}{\phi^2}\right) \times (3\phi - 1) = 8\theta$$

where θ is the reduced temperature

i.e.
$$\theta = \frac{\left(\pi + \dfrac{3}{\phi^2}\right) \times (3\phi - 1)}{8}$$

Substituting the values of π and ϕ,

$$\theta = \frac{\left[0.1096 + \dfrac{3}{(50.607)^2}\right](3 \times 50.607 - 1)}{8}$$

$$= 2.0888$$

We know that reduced temperature

$$\theta = \frac{T}{T_c}$$

$$T = T_c \times \theta$$

Here $T_c = 190.2°K$

Substituting the values,

$$\text{Temperature} = 190.2 \times 2.0888 = 397.3°K.$$

Problem 5:

Calculate the pressure exerted by one mole of carbon dioxide gas in a 1.32 dm³ vessel at 48°C using :

(a) the ideal gas equation and

(b) the van der Waals equation. The van der Waals constants are;

$$a = 3.59\ dm^6\ atm\ mol^{\ 2}$$

and $$b = 0.0427\ dm^3\ mol^{\ 1}.$$

Solution:

(a) For an ideal gas, PV = nRT, so that

$$P = \frac{nRT}{V}$$

$$= \frac{(1\ \text{mol})\ (0.08206\ \text{dm}^3\ \text{atm K}^{-1}\ \text{mol}^{-1})(321\ \text{K})}{1.32\ \text{dm}^3}$$

$$= 19.9\ \text{atm.}$$

(b) For a van der Waals as,

$$P = \frac{nRT}{V - nb} - \frac{n^2a}{V^2}$$

$$= \frac{(1\ \text{mol})\ (0.08206\ \text{dm}^3\ \text{atm K}^{-1}\ \text{mol}^{-1})(321\ \text{K})}{(1.32\ \text{dm}^3) - (1\ \text{mol})(0.0427\ \text{dm}^3\ \text{mol}^{-1})}$$

$$= \frac{(1\ \text{mol})^2 (3.59\ \text{dm}^6\ \text{atm mol}^{-2})}{(1.32\ \text{dm}^3)^2}$$

$$= 2062\ \text{atm} - 2.06\ \text{atm} = 18.56\ \text{atm.}$$

The experimental value of the pressure is 16.40 atm. Thus, van der Waals equation gives a better agreement with experiment than the ideal gas equation.

Problem 6:

Two moles of ammonia are enclosed in a five litre flask at 27°C. Calculate the pressure of ammonia using van der Waals equation. For ammonia,

$$a = 4.17 \text{ atm. litre}^{-2} \text{ mole}^{2}$$

and $$b = 0.0371 \text{ litre mole}^{-1}.$$

Solution:

V is the volume of n (= 2) moles of the gas = 5 litres.

a = 4.17 atms litre^{-2} mole^{-2} (correction due to inter-molecular attraction).

b = 0.0371 litre mole^{-1} (volume correction)

R = 0.082 litre atms. absolute degree^{-1} mole^{-1}.

$$T = 27 + 273 = 300\text{MK}$$

number of moles (n) = 2

According to van der Waals equation,

$$\left(P + \frac{a}{V^2}\right)(V - b) = Rt, \text{ valid for one mole.}$$

$$\left(P + a\frac{n^2}{V^2}\right)(V - nb) = nRT \text{ valid for n moles.}$$

i.e. $$P = \frac{nRT}{V - nb} - \frac{an^2}{V^2}$$

Substituting the values of a, b, R, V and T in equation.

$$P = \frac{2 \times 0.082 \times 300}{5 - 2 \times 0.0371} - \frac{4.17 \times 2 \times 2}{5 \times 5}$$

$$= 9.98 - 0.67 = 9.31 \text{ atm.}$$

Problem 7:

For a fluid show that $(¶P/¶T)_V = a/b$ *where a is expansivity and b is compressibility of the gas. Hence, show that for an ideal gas.*

$$(\partial P/\partial T)_V = R/V$$

Solution:

As a gas can be defined by any two of the three parameters P, V and T, we can write

$$V = f(T \,.\, P)$$

As V is a state function, hence its total differential can be written as follows:

$$dV = \left(\frac{\partial V}{\partial T}\right)_P dT + \left(\frac{\partial V}{\partial P}\right)_T dP \qquad ...(1)$$

By definition, $\alpha = \frac{1}{V}\left(\frac{\partial V}{\partial T}\right)_P$ so that $\left(\frac{\partial V}{\partial T}\right)_P = \alpha V$

and $\beta = -\frac{1}{V}\left(\frac{\partial V}{\partial P}\right)_T$ so that $\left(\frac{\partial V}{\partial P}\right) = -\beta V$

On substituting for α and β in Eq. (i) we get

$$dV = \alpha\, VdT - \beta\, VdP$$

When V is constant, $dV = 0$. Hence,

$$\alpha\, VdT - \beta\, VdP = 0$$

or $$V(\alpha\, dT - \beta\, dP) = 0$$

But $V \neq 0$, hence

$$\alpha\, dT - \beta\, dP = 0$$

or $$\alpha\, dT = \beta\, dP$$

or $$\left(\frac{\partial P}{\partial T}\right)_V = \frac{\alpha}{\beta}$$

For an ideal gas we know, $\alpha = 1/T$ and $\beta = 1/P$. Hence,

$$\left(\frac{\partial P}{\partial T}\right)_V = \frac{\alpha}{\beta} = \frac{1/T}{1/P} = \frac{P}{T} = \frac{R}{V}$$

This is to be proved.

Problem 8:

Calculate the root mean square and average velocity of carbon dioxide at 1000°C.

Solution:

$$(\text{RMSV}) = 1.58 \times \sqrt{\frac{T}{M}} \times 10^4 \text{ cm/sec.}$$

Now $T = 273 + 1000 = 1273$ K; $M = 44$

$$\therefore \qquad (RMSV) = 1.58 \times \sqrt{\frac{1273}{44}} \times 10^4 \text{ cm/sec.}$$

$$= 84870 \text{ cm/sec.}$$

This gives the root mean square velocity.

Now average velocity = 0.9213 × R.M.S.V. of CO_2

Average velocity = 0.9213 × 84870 = 78190 cm/sec.

Problem 9:

Calculate the root mean square velocity of oxygen at N.T.P.

Solution:

For oxygen at N.T.P.

P = 76 cms = 76 × 13.6 × 981 dynes/sq-cm

V = 22.4 litres = 22400 cc.

Substituting the values in the equation:

$$(RMSV) = \sqrt{\frac{3PV}{M}}$$

we have,

$$RMSV = \sqrt{\frac{3 \times 76 \times 13.6 \times 981 \times 22400}{32}}$$

$$= 46140 \text{ cm/sec.}$$

Problem 10:

Calculate the root mean square velocity of nitrogen at 27°C and 700 mm pressure.

Solution:

(i) Conversion of G.M.V. of new volume

At N.T.P.	At New Conditions
P_1 = 760 mm	P_2 = 700 mm
V_1 = 22400 c.c.	V_2 = ?
T_1 = 273 °K	T_2 = 273 + 27 = 300 °K

By gas equation

$$\frac{P_1V_1}{T_1} = \frac{P_2V_2}{T_2}$$

$$V_2 = \frac{P_1V_1}{T_1} \times \frac{T_2}{P_2}$$

$$= \frac{22400 \times 760}{273} \times \frac{300}{700} = 26720 \text{ c.c.}$$

(ii) Calculating the value of RMSV.

Now $P = 700 \text{ mm} = 70 \text{ cm}$

$= 70 \times 13.6 \times 981 \text{ dynes/cm}$

$V = 26720 \text{ c.c.};\ M = 28 \text{ gm}$

Substituting the values in the equation:

$$\text{RMSV} = \sqrt{\frac{3PV}{M}}$$

$$\text{RMSV} = \sqrt{\frac{3 \times 70 \times 13.6 \times 981 \times 26720}{28}}$$

$$= 51710 \text{ c/sec.}$$

5

ARRHENIUS THEORY OF ELECTROLYTIC DISSOCIATION

INTRODUCTION

Arrhenius (1887) successfully put forward a theory of which Clausius's theory was a fore runner. The postulate of this theory known as theory of electrolytic dissociation are as follows:

(a) When dissolve in water, every electrolytic yields two types of charged particles called ions. Those carrying a positive charge are known as cations, while those carrying negative charge are known as anions.

However, according to the modern view, a solid electrolyte is a combination of charged ions, bound by electrostatic lines of forces. When dissolved in solution, these lines of forces are cut and the ions separate.

$$\underset{}{AB} \rightarrow \underset{\text{caution}}{A^+} + \underset{\text{anion}}{B^-} \quad \text{(old view)}$$

$$A + B^- \rightarrow A^+ + B^- \quad \text{(Modern view)}$$

(b) Ions present in solution recombine to form unionised electrolyte and there is a state of equilibrium to which the law of mass action can be applied.

Therefore, $A^+B^- \rightleftharpoons A^+ + B^-$

According to the law of mass action,

$$K = \frac{[A^+][B^-]}{[A^+B^-]}$$

where K is known as ionisation or dissociation constant. The fraction of the total number of molecules present as ions in solution is known as degree of dissociation.

(c) Each ion that is formed as a result of electrolytic dissociation produce the same effect on colligative properties (like osmotic pressure, lowering of vapour pressure, elevation of boiling point etc.) as an undissociation molecule. In simpler words, each ions behaves osmotically as a molecule. For example, an electrolyte yielding there ions per molecule will exert thrice the normal effect of the electrolyte, had the electrolyte not dissociated.

(d) On the passage of electricity, cautions and anions move towards the cathode and anode, respectively. This movement of ions towards the electrodes constitutes the electric current through the solution. This explains the process of electrolysis.

(e) The properties of the electrolytes in solution are the properties of the ions. All cobalt salt solution are pink, because each solution contains cobalt ions which are pink in colour. Similarly, all chromium salt solution are green, as chromium ions are green.

Facts explained by Arrhenius theory-Arrhenius theory could successfully explain all the phenomena concerning electrolytes. It explains the anomaly of the electrolytes towards the van't Hoff theory of dilute solutions. Besides these, the degree of dissociation of electrolytes calculated on the basis of this theory conform well with the values obtained by the other methods. All these facts led to the universal acceptance of this theory.

(a) *Explanation of Electrolysis :* On assuming the presence of ions in an electrolytic solution, the conduction of electricity through them can be explained. When a difference of potential is applied across two metal electrodes dipping in the solution of an electrolyte, the cations and anions move gain of electrons, as shown below:

$$A^+B^- \rightleftharpoons A^+ + B^-$$

$$B^- \rightarrow B + e$$

$$A^+ + e^- \rightarrow A$$

The electron liberated at the anode travels from the anode the cathode outside the cell, through the connecting wire. This electron is now available for the conversion of A^+ ion at the cathode. The circuit is completed and the electricity flows through the solution. These ions after losing their charges are liberated as atoms as molecules. This explains the chemical reaction caused due to the passage of electricity. This process is known as electrolysis. Solutions of non-electrolytes do not produce ions, hence they are unable to conduct electricity.

(b) *Variation of Equivalent Conductivity with Dilution* : As the degree of dissociation of an electrolyte increases on dilution, therefore, we can easily explain the increase in equivalent conductivity on dilution, as it depends upon the number of ions. At a certain dilution known as infinite dilution, equivalent conductivity acquires a maximum value. This can be explained on the basis that at a certain dilution, the dissociation of an electrolytes is complete. In other word, the number of ions produced is maximum and does not increase further on dilution. Hence, a constant maximum value of equivalent conductivity is reached at infinite dilution.

(c) *Heat of Neutralisation* : We know that the heat of neutralisation of strong acids and strong bases is about 13.7 k cals. This fact can also be explained on the basis of electrolytic dissociation. As strong acids and bases are almost completely ionised, the process of neutralisation can be represented as:

$$\underbrace{H^+ + A^-}_{\text{from acid}} + \underbrace{B^+ + OH^-}_{\text{from base}} \rightarrow \underbrace{A^- + B^+}_{\text{from electrolyte}} + H_2O + 13.7\text{ k cals}$$

or $H^+ + OH^- \rightarrow H_2O + 13.7\text{ k cals}$

Thus, the heat of neutralisation is simply the heat of formation of water from its ions and since this reaction is involved in all neutralisation processes, the heat evolved is always the same. In the case of a weak acid or weak bases or both the heat neutralisation is less than 13.7 k cals. This is because the weak substance is dissociated for complete neutralisation and the dissociation process requires energy and is therefore, endothermic. The energy is taken from the system itself. Thus, the net heat of neutralisation envolved is less than 13.7 k cals.

(d) *Common ion Properties* : We know that solutions of electrolytes with a common ion show similar physical and chemical properties. Thus, we know that Cu^{+2}, MnO_4^-, CO^{+2} ions are blue, purple, pink, respectively. Thus, any solution of the salt containing these ions will possess the same colour and will also have the same absorption spectrum. Similarly, all cyanides are poisonous, as CN^- ions are poisonous in nature.

(e) *Ionic Reaction :* Evidence for the existence of ions in aqueous solutions can be had by a number of reactions in inorganic chemistry. In qualitative analysis, the radicals are tested as long as they are in the form of ions. Thus, Ag^+ ion can be tested if it is present in the solution, by adding HCl solution.

$$AgNO_3 \rightarrow Ag^+ + NO_3^-$$

$$HCl \rightarrow H^+ + Cl^-$$

$$Ag^+ + Cl^- \rightarrow AgCl \text{ (white ppt.)}$$

If silver ions do no exist in the solution, we will not get a white precipitate on adding HCl solution. Thus, when one of the ions combines with a molecule to form a complex ion, the solution does not give the test of that ion. If silver exists as $NaAg(CN)_2$, then no test no test for Ag^+ ions is obtained as the compound gives only Na^+ and $Ag(CN)_2^-$ ions on dissociation.

Similarly, an acid which gives all the tests for hydrogen ion when dissolved in water, does not give the same tests when dissolved in organic solvents. This is because the acid does not ionise in organic solvents to give the hydrogen ions.

(f) *Colligative Properties of Electrolytic Solutions in Water :* On the basis of Arrhenius theory, one can explain the anomalous behaviour of electrolytes in aqueous solution. van't Hoff showed that colligative properties like osmotic pressure, lowering of vapour pressure etc. for electrolytes were higher than those for non-electrolytes. It was seen that the elevation in boiling point due to the presence of one gram mole of an electrolyte dissolved in 1000 gm of water for a non electrolytes (like cane sugar, urea etc.) was 0.52°, while that for electrolytes like NaCl, HNO_3, KCl was found of approach a value of 1.34° (twice that of 0.52°) at large dilutions. Similarly, for electrolytes like Na_2SO_4, $BaCl_2$ etc. the value was about 1.56° (thrice that of 0.52°) at large dilutions. In order to explain this anomaly, van't Hoff introduced a new factor known as van't Hoff factor (i), which is defined by,

$$i = \frac{\text{Observed elevation of boiling point}}{\text{Theoretical elevation of boiling point}}$$

In general, we can say that

$$i = \frac{\text{Observed colligative property of an electrolyte}}{\text{Theoretical colligative property of an electrolyte}}$$

Factors influencing ionisation—

(a) *Nature of Solute :* Strong electrolytes are completely ionised in solution and such electrolytes are obtained by the interaction between strong acids and weak bases are not so strongly ionised.

(b) *Nature of Solvent :* The solvent cuts the lines of forces binding the two ions and hence breaks them. Greater the capacity of the solvent, greater will be the ionisation of the electrolyte.

This tendency of solvent is measured in terms of dielectric constant, which is defined as, 'the capacity to weaken the forces of attraction, between the electrical charges immersed in that solvent.'

(c) *Temperature :* The degree of ionisation of an electrolyte in solution is proportional to the temperature. At higher temperatures, the increased molecular velocities cut the forces of attraction between the ions to a greater extent and thus cause increased ionisation.

(d) *Concentration :* The degree of ionisation of an electrolytes is inversely proportional to the concentration. This is due to the fact that at higher dilution, the solvent molecules make more molecules of the solute to break in to ions.

On substituting the value of λ_v, λ_∞ and V, the value of K_a can be calculated. At 20°C the value of K_a for acetic acid is found to be 1.8×10^{-5}.

Dissociation Constants of Phosphoric Acid

Phosphoric acid (H_3PO_4) is a tribasic acid and it ionises in three stages and consequently it has *three* dissociation constants, one each for each stage. Thus

$$H_3PO_4 \overset{K_1}{\rightleftharpoons} H^+ + H_2PO_4^- \quad \text{...(i)}$$

$$H_2PO_4^- \overset{K_2}{\rightleftharpoons} H^+ + HPO_4^{2-} \quad \text{...(ii)}$$

$$HPO_4^{2-} \overset{K_3}{\rightleftharpoons} H^+ + PO_4^{3-} \quad \text{...(iii)}$$

$$\because \quad K_1 = \frac{[H^+][H_2PO_4^-]}{[HPO_4^{3-}]};$$

$$K_2 = \frac{[H^+][HPO_4^{2-}]}{[H_2PO_4^-]}$$

and $$K_3 = \frac{[H^+][PO_4^{3-}]}{[HPO_4^{3-}]}.$$

The dissociation of acids (i), (ii) and (iii) into H^+ ions becomes more and more difficult due to increasing negative charge on the acid molecule. Hence $K_1 > K_2 > K_3$. The experimental values are:

$$K_1 = 1.1 \times 10^{-2};\ K_2 = 2 \times 10^{-7},$$

and $$K_3 = 3.6 \times 10^{-12}.$$

EVIDENCES IN FAVOUR OF THE THEORY

(i) *Validity of Ohm's Law (E = RC) in Electrolytic Solutions and Hence Explanation of Electrolysis:* The electrolysis starts even when a current of very small e.m.f. is applied. This can happen *only if ions are already present before electrolysis.* This fact is also supported by the fact that electrolytes obey Ohm's law which go to show that *whole of the current passed is utilized only in overcoming the resistance of the ions to move* and no part of it is utilized in breaking the molecules into ions. Thus, on passing electric current during electrolysis, the ions already present in solution move towards opposite electrodes and on reaching the respective electrodes they are discharged and are either deposited there or evolved as a gas or undergo secondary changes.

(ii) Heat of Neutralization: Whenever one gram equivalent of a strong acid (like HCl, H_2SO_4, HNO_3, etc.) in solution is neutralized with 1 gram equivalent of a strong base (like NaOH, KOH, etc.) in solution, the same quantity of heat, viz. 13,700 calories is produced. This is explained on the basis of ionic theory that strong acid and strong bases are nearly completely ionized in solution and the heat of neutralization is nothing but the heat of formation of water molecules from H^+ ions (of acid) and OH^- ions (of base). Hence, this value is always the same irrespective of the nature of the strong acid and base used.

$$(H^+ + Cl^-) + (Na^+ + OH^-) = (Na^+ + Cl^-) + H_2O + 13{,}700 \text{ cals}$$

or $$H^+ + OH^- = H_2O + 13{,}700 \text{ cals}.$$

However, the heat of neutralization of a weak acid or weak base is less. For example, when one 1 gram equivalent of acetic acid is neutralized by 1 gram equivalent of NaOH, 13,400 cals. of heat are evolved. This difference (13,700 – 13,400 = 300 cals) is nothing but the *heat of dissociation of weak acid, i.e.,* the heat required to dissociate the weak acid, acetic acid, which is not fully dissociated.

(iii) *Presence of Ions in the Solid Salts:* X-ray diffraction studies have shown that ions are also present in solid salts. Thus, a crystal of sodium chloride does not contain NaCl units, but it consists of each Cl^- ion surrounded by six Na^+ ions and each Na^+ ion surrounded by six Cl^- ions. *The solid crystal is not a good conductor, because the crystal lattice restricts the movement of ions.* However, when such a crystal is dissolved in a suitable solvent, the attractive forces between the ions weaken and the crystals fall apart into free Na^+ and Cl^-, which can move freely under the influence of electricity and hence the salt solution becomes good conductor.

(iv) *Colour of Solution :* Certain compounds possess colour in solution. *The colour of solution is due to the presence of coloured ions.* Thus, Cu^{2+} ions are blue and any solution of its salt like $Cu(NO_3)_2$ or $CuSO_4$ will always show blue colour on *sufficient dilution.* Similarly, CO^{2+} ions and Fe^{2+} ions are pink and yellow coloured respectively and their presence impart these colours to solution.

(v) *Ionic Reactions:* The reactions of electrovalent compounds are nothing but the reactions of ions in solution. Thus, the formation of white precipitate of AgCl, when NaCl solution is reacted with $AgNO_3$ solution is nothing but the reaction of Ag^+ ions and Cl^- ions.

$$(Na^+ + Cl^-) + (Ag^+ + NO_3^-) = (Na^+ + NO_3^-) + AgCl\downarrow$$

or

$$Cl^- + Ag^+ \rightarrow AgCl\downarrow$$

A solution of $FeCl_3$ gives a brown precipitate of ferric hydroxide; while a solution of $K_3Fe(CN)_6$ does not give a brown precipitate on addition of NH_4OH, because the latter does not yield Fe^{3+} ions which may combine with OH^- ions to form precipitate of $Fe(OH)_3$.

(vi) *Strength of Electrolytes:* On the basis of extent of ionization the *compounds such as* HCl, NaOH, NaCl, *etc., which are almost fully ionized in water, are called strong electrolytes; while those* (*like* CH_3COOH, NH_4OH), *which have low degree of ionization, are called weak electrolytes* However, these terms are relative as the degree of ionization varies in different solvents. Thus, HCl is a strong electrolyte in aqueous solution, but it is very weak electrolyte in non-aqueous solvents like CCl_4. On the other hand, NH_4OH is a weak electrolyte in aqueous solution, but will be strong electrolyte in acetic acid medium.

(vii) *Colligative Properties:* Colligative properties like osmotic pressure, depression in freezing point, elevation in boiling point and lowering of vapour pressure are a function of the number of particles (molecule as well as ions) of the solute in solution. Thus, a decimolar solution of urea has been found to extent the same osmotic pressure as a decimolar solution of sugar. On the other hand, a decimolar solution f NaCl under the same conditions gives approximately the double value of osmotic pressure This clearly shows that the *number of particles present in HCl solution is approximately double the number of particles present in solution of urea or sugar*. This is explained by the fact that urea and sugar, being non-electrolyte, do not dissociate; while HCl, being an electrolyte, undergoes ionization to give approximately double number of particles (HO^+, Cl^- ions). Similarly, the abnormal values of depression in freezing point, elevation in boiling point in case of electrolytes can only be explained on the basis of theory of ionization.

(viii) *Heat of Precipitation:* The heat of precipitation of AgCl formed by the addition of 1 g equivalent of any soluble chloride to 1 g equivalent of Ag salt solution is always the same. This is explained on the basis of ionic theory according to which the heat of precipitation of AgCl is nothing but the heat of combination of Ag^+ ions and Cl^- ions to give AgCl precipitate

$$Ag^+ + Cl^- \rightarrow AgCl + x \text{ cals.}$$

(ix) *Increase of Conductivity on Dilution:* The conductivity of a solution is due to the presence of ions formed by dissociation. As we know, with the increase of dilution, the degree of ionization of electrolyte and the total number of ions in solution increases and heat conductivity also increases.

(x) *Ostwald's Dilution Law:* The Ostwald's dilution law is based on the theory organization applied to weak electrolytes. The validity of Ostwald's dilution formula in weak electrolytes is most reliable proof of the dissociation fo molecules into ions.

(xi) When a tube containing NaI is whirlled very rapidly, the farthest end is found be + very charged. This is explained on the basis of ionic theory that NaI is present more as Na^+ and I^- ions and the lighter Na^+ ions move more rapidly to the ends as compared heavy I^- ions.

INTRODUCTION OF ELECTROLYSIS

The process of decomposition of an electrolyte by the passes of electric current through it is known as electrolysis. For bringing about electrolysis, we need a vessel known as 'electrolysis vessel'. In this vessel, an electrolytic solution is taken and two metallic rods are dipped in it. These two rods are then connected to the two terminals, (*i.e.*, + and –) of the battery. These metallic rods which lead the electric current to and form the electrolyte are called electrodes. The electrode connected with the positive end of the battery is called the positive electrode or anode, whereas the rod connected to the negative end of the battery is known as negative electrode or cathode. This complete arrangement is known as electrolytic cell. The current enters the electrolyte through the anode and leaves through the cathode. The electrically charged atoms or radical that move to the electrodes are called ions. Ion travelling towards the cathode is known as cation and ion travelling towards anode is known as anion. The liquid surrounding the cathode is known as catholyte and that round the anode as *anolyte*.

For practical purpose, the international values are taken to be identical with the absolute value.

For steadily, direct currents, we have Ohm's law given by E=RC

i.e. the emf., E of the circuit measured in volts, is equal to the product of the resistance, R in ohms and current, C in amperes. Ohm's law is not only applicable to a complete electric circuit but also to any part of the circuit.

MOVEMENT OF IONS DURING ELECTROLYSIS

Who have already said that the flow of electricity in an aqueous solution results due to the movement of ions.

When an electric field is applied to electricity in an aqueous solution results due to the movement of ions.

When an electric field is applied to electrolytic solution, the movement of ions occurs towards respective electrodes.

A filter paper strip (about 3 × 8 cm) is cut out. It is moistened with a little dilute potassium nitrate solution. Then, it is kept on a clean watch glass. A battery of 15 volts is connected across the filter paper by using a pair of crocodile clips. At the centre of filter paper strip, a suspension of yellow copper chromate in potassium nitrate is placed. When the electric current is applied across the filter paper strip, two zones are observed on the filter paper. A blue zone is seen moving towards the cathode which is due to the migration of Cu^{2+} ions. A yellow zone due to chromate ion, CrO_4^{-2}(aq) is seen moving towards the anode.

The above experiment can also be done with coloured, water-soluble ionic compounds such as potassium permanganate, cobalt chloride, nickel chloride, etc.

Resistance or Conductance

Metallic conductors as well as electrolytes obey Ohm's law, which states that "*The strength of current (I) flowing through a conductor is directly proportional to the potential difference (E) applied across the conductor and inversely proportional to the resistance (R) of the conductor.*"

Thus, the mathematical form of Ohm's law becomes as

$$I = E/R.$$

In relation to electrolytes, the term conductance (C) is used more frequently than resistance. Conductance implies the ease with which electric current can flow through a conductor. It is therefore, defined as the reciprocal of resistance,

Conductance = 1/R.

Units: It is expressed in units of reciprocal or inverse ohms, $ohms^{-1}$ or mhos.

Resistance

The resistance R of a conductor is

(i) directly proportional to its length (l cm) and

(ii) inversely proportional to its area of cross of cross section.

Hence $R \propto \frac{1}{a}$ or $R = \rho \frac{1}{a}$ ...(1)

where ρ is a constant depending on the nature of the material of the conductor, and is called the "*specific resistance*" or "*resistivity.*"

If l = 1 cm and a = l sq. cm, then ρ = R ohms.

Hence, specific resistance may be defined as:

"*The resistance of a uniform column of the material of the conductor having a length of 1 cm and a cross section of 1 sq. cm.*"

Units: As we know that

$$\rho = R\frac{a}{l} = \frac{\text{Ohm (cm)}^2}{\text{cm}} = \text{ohm. cm} \quad ...(2)$$

SPECIFIC CONDUCTANCE

The specific conductance of a conductor is the reciprocal of specific resistance, and is generally denoted by K (Greek, Small Kappa). Equation (2) can be written as

$$R = \frac{1}{K}\frac{1}{a} \text{ ohms} \qquad \left[\therefore \rho = \frac{1}{K}\right] ...(3)$$

$$K = \frac{1}{R} \times \frac{1}{a} \text{ ohms}^{-1} \text{ com}^{-1}$$

If l = 1, and a = 1 sq. cm. equation (3) becomes as

$$K = \frac{1}{R} = \rho$$

In the case of solution, the definition will be the same as of a conductor. In simple words, it is defined as:

"*The conductivity offered by a solution of length 1 cm area of unit cross section.*"

Units: We know that K = 1/ρ

Hence the unit of K will be

$$\frac{1}{\text{ohm. cm.}} = \text{ohm}^{-1} \text{ cm}^{-1} = \text{mho/cm.}$$

Equivalent Conductance

It may be defined as,

"*The conductance of a solution containing 1 gm. equivalent of an electrolyte when placed between two sufficiently large electrodes which are 1 cm apart.*"

It is denoted by λ_r, where V is the volume in c.c. containing one gram equivalent of electrolyte dissolved in it and is measured in reciprocal ohm, or mho.

MOLECULAR CONDUCTANCE

It is defined as,

"*The conductance of a solution containing 1 gm mole of the electrolyte when placed between two sufficiently large electrodes placed one cm apart.*"

It is denoted by symbol μ, and is measured in mhos,

Relation Between Specific Conductance and Equivalent Conductance

Consider a rectangular metallic vessel with opposite sides exactly 1 cm apart. If now 1 c.c. of solution is placed in this vessel, the area of the opposite faces of the cube converted by the solution will be 1 sq. cm. The measured conductance of the solution will be its specific conductance. If 1 c.c. of the solution is placed in the above vessel containing 1 gm. equivalent of the electrolyte, then by definition, the measured conductance will also be equal to the equivalent conductance,

$$\lambda = K$$

Now if 9 c.c. of pure solvent are added so that the total volume now occupied by the solution is 10 c.c.; the (because 10 c.c. of the diluted solution still contains 1 gm. equivalent of the electrolyte). But by definition of the specific conductance, the measured conductance of the diluted solution will be ten times the specific conductance,

$$\therefore \qquad \lambda = K \times 10$$

Similarly if the solution is diluted 100 times its volume, the measured conductance will still be equivalent conductance, but will be 100 times the specific conductance.

Thus, $$\lambda_v = K \times 100$$

In general, if the solution contains 1 gm equivalent of the electrolyte dissolved in V c.c. of the solution, then

$$\lambda_r = K \times V$$

∵ Equivalent conductance = specific conductance × volume of solution in c.c. containing 1 gm. equivalent of the electrolyte.

Relation between Molecular conductance and Specific Conductance: As discussed above, a similar type of the relation also exists between molecular conductance and specific conductance *i.e.*,

$$\mu_r = K \times V$$

or Molecular conductance = [Specific conductance]
× [Volume of solution in c.c. containing
1 gm. mole of electrolyte]

Effect of Dilution

(i) *Conductance:* The conductance of s solution is due to the presence of ions in solution. Since on dilution the degree of dissociation of electrolyte in increased and more ions are produced in solution, it is thus expected that the conductance of a solution increase on dilution.

(ii) *Specific Conductance:* Specific conductance depends upon the number of ions present per c.c. of the solution. Since on dilution the degree of dissociation increases but the number of ions per c.c. decreases, it is therefore expected that the specific conductance of a solution should decrease on dilution.

(iii) *Equivalent Conductance:* The value of equivalent conductance increases on dilution. This increase is due to the fact that equivalent conductance is the product of specific conductance and the volume V of the solution containing 1 gm equivalent of the electrolyte.

Since the decreasing value of specific conductance is more than compensated by the increasing value of V, the value of λ_y will increase with dilution.

(iv) *Molecular Conductance: As* discussed in equivalent conductance, the molar conductance of an electrolyte increases with dilution.

(v) The variation of equivalent conductance, with dilution where the equivalent conductance of different dilutions is plotted against $\sqrt{(c)}$, where c is the concentration of the solution in moles per litre.

It is evident that the equivalent conductance tends to become maximum as the concentration decreases or dilution increases.

However the maximum conductivity in case of strong electrolytes like KCl is reached at relatively higher concentrations (or lower dilutions) that the case of weak electrolytes like CH_3COOH.

EQUIVALENT CONDUCTANCE AT INFINITE DILUTION

The maximum value of the equivalent conductance of an electrolyte obtained at high dilutions is known as the equivalent conductance at infinite dilution. It is denoted by λ_∞. By infinite dilution we mean that the solution is so dilute that any further dilution does not produce any charge in the equivalent conductance of the solution.

Measurement of Specific Conductance

We know $$R = \rho \times \frac{l}{a}$$

$$\frac{1}{\text{obs. conductance}} = \frac{1}{\text{sp. conductance}} \times \frac{l}{a}$$

$$\text{Sp. conductance} = \text{Obs. conductance} \times \frac{l}{a}$$

$$= \text{Obs. conductance} \times \text{cell constant.}$$

The ratio *l/a* is known as *cell constant* and is generally denoted by *x*. It depends upon the area of cross section of electrodes and the distance between them. Thus, the determination involves the measurement of conductance of the solution and of the cell constant.

Measurement of Conductance: The conductance of a solution can be expressed by

$$\frac{1}{R} = x(c_1\lambda_1 + c_2\lambda_2 + c_3\lambda_3 \ldots + c_1\lambda_1)$$

Here c_i terms are the concentrations of various ions in solution, λ_i terms are numerical constants characteristic of ions and x is a proportionality factor that takes account of the geometry of the cell. Because of interionic interactions and solvent-ion interactions, the values of λ_i depend slightly on concentration, and extrapolation to infinite dilution is required to obtain values that are physical constants of ions. Thus with proper calibration, a single measurement can give the concentration of any electrolyte.

The conductance of a solution can be measured by finding out the resistance by means of a Whetstone bridge arrangement. However when D.C. current is used in the solutions, the following difficulties may arise:

(i) During the conduction of D.C. current, there starts electrolysis which results in a change in the concentration of the solution.

(ii) Products of electrolysis get accumulated at the electrodes and set up a back E.M.F. Above mentioned difficulties are overcome by using A.C. current in the following manner.

(i) *The polarisation produced due to A.C. current in one direction is neutralised by the effect of current in other direction.*

(ii) *No electrolysis takes place.*

When an A.C. current is used, the following modifications are made:

(i) *An A.C. current of frequency fo 1000-4000 cycles per second is generally used, an oscillator has to be used.*

(ii) *An alternating current cannot be employed to detect the null-point and is therefore replaced by a Head phone.*

(ii) *In order to minimise the polarisation still further, the platinum electrodes in the cell are coated with finely divided platinum. This gives rise to a larger surface to the electrodes and consequently the polarisation is retarded.*

Method : The apparatus used for the purpose is a modified form of the Wheatstone bridge.

(i) The solution whose conductance is to be measured is taken in the conductivity cell. The conductivity cell consists of electrodes which are platinum coated with finely divided platinum. These are welded to platinum wires which are sealed through two glass tubes. The tubes are firmly fixed in an ebonite cover so that the distance between the electrodes is not altered during the experiment. The copper wires of the circuit dip in mercury placed in glass tubes to complete the circuit. Some other types of conductivity cells.

(ii) The cell is connected to a resistance box R on one side and to a long thin wire AB stretched along a scale on the other side.

(iii) Some known resistance 'R' is taken out of the resistance box. An alternating current of about 1000 - 4000 cycles per second is passed through the solution. The sliding contact d is moved on the wire AB so that minimum or no sound is heard in the head phone. Then,

$$\frac{\text{Resistance of the solution}}{\text{Resistance R}} = \frac{\text{Length Bd}}{\text{Length Ad}}$$

Thus, knowing R, Ad and Bd the resistance of the solution can be calculated.

$$\therefore \text{Conductivity of the solution} = \frac{1}{\text{Resistance of the solution}}$$

Determination of Cell Constant: The cell constant is determined by substituting the value of sp. conductance of N/50 KCl solution at 25°C as accurately determined by Kohlrausch (0.002765) and the value of conductance is observed with the given cell in the relationship:

$$\text{Cell constant} = \frac{0.002765}{\text{obs. conductance}}$$

Measurement of Equivalent Conductance: Knowing specific conductance as above, the equivalent conductance of the solution can be measured by measuring the specific conductance K with volume V in c.c. containing 1 gm. equivalent of the electrolyte

$$\lambda_r = K \times V.$$

KOHLRAUSCH'S LAW

The values of equivalent conductances at infinite dilution for certain pairs of electrolytes at 18°C are given in Table 1.

Table 1

Pairs of electrolytes with the same anion	λ_0 *Ohm*$^{-1}$	*Difference Ohm*$^{-1}$	*Pairs of electrolytes with the same cation*	λ_0 *Ohm*$^{-1}$	*Difference Ohm*$^{-1}$
KF	111.2		KCl	130.0	
		21.1			18.8
NaF	90.1		KF	111.2	
KCl	130.0		NaCl	108.9	
		21.1			18.8
NaCl	108.9		NaF	90.1	

From the above table it follows that

(i) When potassium is replaced by sodium in any of the electrolytes, there is always a difference of 21.1 ohm^{-1} in equivalent conductances at infinite dilution irrespective of the nature of the anion with which they are associated.

(ii) When chlorine is replaced by fluorine, there is always a difference of 18.8 ohm^{-1} in equivalent conductances at infinite dilution irrespective of the nature of the cation with which they are associated. The above observations led Kohlrausch (1875) to formulate a law known after his name as Kolhrausch's law. This law can be stated as follows:

"*The equivalent conductance at infinite dilution for any electrolyte is the sum of equivalent conductances for the cation and anion at infinite dilution.*"

Mathematically, Kohlrausch's law can be put as

$$\lambda_\infty = \lambda_\infty^+ + \lambda_\infty^-$$

where λ_∞^+ and λ_∞^- are the equivalent conductances of the cation and anion respectively at infinite dilution.

APPLICATIONS OF KOHRAUSCH'S LAW

(i) *Calculation of Equivalent Conductance at Infinite Dilution for Weak Electrolytes:* Equivalent conductance at infinite dilution can be calculated by a graphical method. But the graphical method is not applicable to weak electrolytes as they are very slightly ionized. But Kohlrausch's law can be used to determine the value of λ_∞ for weak electrolytes. For example, equivalent conductance of acetic acid at infinite dilution can be calculated from the equivalent conductances at infinite dilution of hydrochloric acid, sodium acetate and sodium chloride as follows:

$$\lambda_\infty HCl = \lambda_\infty H^+ + \lambda_\infty Cl^- \quad \text{(say x)} \quad ...(1)$$

$$\lambda_\infty CH_3COONa = \lambda_\infty Na+ + \lambda_\infty CH_3COO^- \quad \text{(say y)} \quad ...(2)$$

$$\lambda_\infty NaCl = \lambda_\infty Na^+ + \lambda_\infty Cl^- \quad \text{(say z)} \quad ...(3)$$

Add equations (1) and (2) and subtract (3), we get

$$\lambda_\infty HCl + \lambda_\infty CH_3COONa — \lambda_\infty NaCl$$

$$= \lambda_\infty H^+ + \lambda_\infty Cl^- + \lambda_\infty Na^+ + \lambda_\infty CH_3COO^- — \lambda_\infty Na^+ - \lambda_\infty Cl^-$$

$$x + y - z = \lambda_\infty H^+ + \lambda_\infty CH_3COO^-$$

$$= \lambda_\infty CH_3COOH$$

or $\lambda_\infty CH_3COOH = x + y - z.$

Since the values of *x*, *y* and *z* are experimentally determined, we can easily calculate the value of λ_∞ for acetic acid.

(ii) *Calculation of Degree of Dissociation of Weak Electrolytes* : The value of λ_0 helps to calculate the degree of dissociation and the dissociation constants of weak acids and weak bases.

For example, the degree of dissociation α_c of a weak electrolyte at the concentration c equivalents per litre may be given by the following relation:

$$\alpha_c = \frac{\lambda_c}{\lambda_0}$$

where λ_c is the equivalent conductance of electrolyte a$^+$ concentration c and λ_0 is the equivalent conductance of the same electrolyte at infinite dilution. Hence, measurement of λ_c permits evaluation of α_c if λ_0 known.

(iii) *Determination of Solubility of Springly Soluble Salts* : For ordinary purposes ionic salts like AgCl, $BaSO_4$, $CaCO_3$, Ag_2CrO_4, $PbSO_4$, PbS, $Fe(OH)_3$, etc. are regarded as insoluble, but they do have a *very small* but *definite* solubility in water.

The solubility of such sparingly soluble salts is obtained by determining the specific conductivity (K) of a saturated salt solution. Since only a very small amount of salt is present in solution. The equivalent conductivity at such high dilution can practically be taken as λ_∞, *i.e.*,

For very sparingly soluble salts; $\lambda_r = \lambda_\infty = K.V$

where V is the volume in cm3 containing 1 g equivalent of salt.

The value of λ_∞ can be calculated with the help of Kohlrausch's law *i.e.*,

$$\lambda_\infty = \lambda_+ + \lambda_-$$

Thus, knowing the values of K and λ_∞, the value of V can be calculated:

$$V = \frac{\lambda_r}{K} \simeq \frac{\lambda_\infty}{K} = \frac{\lambda_+ + \lambda_-}{K}$$

But V cm^3 of saturated solution contains = 1 g eqt. of salt

$\therefore$ 1000 cm^3 (or litre) = solution contains = 1000/V g eqt. of salt

Hence, solubility of salt,

$$S = \frac{1000}{V} \text{ g equivalent/litre}$$

$$= \frac{1000\,E}{V} \text{ g/litre} = \frac{1000 \times E \times K}{(\lambda_+ + \lambda_-)} \text{ g/litre}$$

where E is the equivalent weight of salt.

APPLICATION OF CONDUCTIVITY MEASUREMENTS

(a) Basicity of Organic Acids : From an examination of the conductances of the sodium salts of a number of organic acids, Ostwald (1887) discovered the empirical relation

$$\lambda_{1024} - \lambda_{32} = 10.8b \qquad ...(1)$$

where λ_{1024} and λ_{32} are equivalent conductances of the salt at 25°C at dilutions of 1024 and 32 litres per equivalent, respectively and b is the basicity of the acid.

Procedure : To prepare N/32 solution of the sodium salt of an acid, a N/16 NaOH solution is accurately prepared. A known volume of it, say 50 ml is neutralised by adding dropwise a concentrated solution of the acid, using an indicator paper. When the neutralisation is complete, the volume is made upto 100 ml. This gives N/32 solution of the salt and the N/1024 solution is prepared by a careful dilution of the solution.

Measure the resistance of N/32 and N/1024 solution of the taking all the usual precautions. Using cell constant, calculate the equivalent conductances at both the dilutions. Then, calculate the basicity of the acid using the formula given above.

Limitations

(i) This rule applies to some organic acids but does not succeed with inorganic acids,

(ii) This method fails when applied to very weak acids whose salts are considerably hydrolysed in solution.

(b) Degree of Hydrolysis : When the salt of a weak acid or base is dissolved in water, hydrolysis occurs, so that the conductance of the solution is now partly due to the ions of the salt and partly due to the ions of the salt and partly due the ions (H^+ or OH^-) of the acid or base formed by hydrolysis. An excess of the weak acid or base in the presence of its salts can be regarded as completely unionised. Therefore, addition of weak acid or base to the salt solution suppresses the hydrolysis but the ionisation will be unaffected.

Aniline hydrochloride ($C6H_5NH_3^+Cl^-$), the salt of a weak base and a strong acid, is hydrolysed in water giving (1 – α) equivalent of unhydrolysed salt and α equivalents of free acid and free base, where α is the degree of hydrolysis.

$$\underset{1-\alpha}{C_6H_5NH_3^+} \rightleftharpoons \underset{\alpha}{C_6H_5NH_2} + \underset{\alpha}{H^+}$$

The weak undissociated base aniline does not contribute to the total equivalent conductance λ of the solution, which may, therefore, be written as

$$\lambda = (1 - \alpha)\lambda_c + \alpha\lambda_{HCl}$$

or

$$\alpha = \frac{\lambda - \lambda_c}{\lambda_{HCl} = \lambda_c} \quad ...(2)$$

which λ_c and λ_{HCl} are the equivalent conductances of the unhydrolyzed salt and free acid at the same concentration. λ_c as has been pointed out, can be determined by measuring the equivalent conductance in the presence of excess aniline.

The value of λ_{HCl} at concentrations M/32, M/64, M/128 are known to be respectively 393, 399 and 401. Thus α being determined from eq. (2) the hydrolysis constant K_h is easily obtained from the following relation:

$$K_h = \frac{\alpha^2 C}{1 - \alpha}$$

(c) Solubility of a Sparingly Soluble Salt : This is based on the simple fact that the saturated solution of a sparingly soluble salt is so dilute that the electrolyte present there in may be regarded as completely ionised. The relationship between concentration C (in gm. equivalents per litre), the equivalent conductance λ, and the specific conductance K, is given by the equation

$$\lambda = K \times \frac{1000}{C} \quad ...(3)$$

The solution being saturated, C represents the solubility (in gram equivalents per litre). Since the solution (though saturated with respect to the almost insoluble salt) is so dilute, it may be regarded as being practically ionised at infinite dilution. Hence λ may be put equal to λ_∞.

This as we know, is equal to $\lambda_{(cation)} + \lambda_{(anion)}$, the figures for which are available from published tables. Hence the solubility of the sparingly soluble salt, C, is

$$C = \frac{100\ K}{\lambda_{(cation)} + \lambda_{(anion)}} \text{ gm equivalent/1000 ml} \qquad ...(4)$$

So, the only experimental work required is to determine the conductance of a saturated solution.

(d) Ionic Product of Water, K_w : This can be calculated from the specific conductance of pure water, as determined experimentally, and its equivalent conductance, as calculated from the λ_0 values of its two ions. The equivalent conductance at infinite dilution λ_∞ is defined by the equation

$$\lambda_\infty = 100 \frac{K_\infty}{C} \text{ mhos cm}^2 \qquad ...(5)$$

where K_∞ is the specific conductance at infinite dilution and C the concentration in gm. equivalent per litre. The most accurate experimental value of specific conductance of purest water so far available is 5.54×10^{-8} mho cm^{-1} at 25°C. This may be put equal to K as pure water may be regarded as a solution of ionised water at infinite dilution.

The values of equivalent conductances of the two ions of water at infinite dilution are:

$$\lambda_\infty(H^+) = 349.8 \text{ mhos cm}^2$$

$$\text{and } \lambda_\infty(OH^-) = 198.6 \text{ mhos cm}^2$$

Hence the equivalent conductance at infinite dilution, λ_∞, of water may be equal to 349.8 + 198.6 *i.e.* 548.4 mhos cm^2. Substituting the values of K_∞ and λ_∞ in equation (5), we get value of C for each of the ions of water:

$$[H^+] = \frac{1000 \times 5.54 \times 10^{-4}}{548.4} = 1.01 \times 10^{-7}$$

$$[OH^-] = \frac{1000 \times 5.54 \times 10^{-4}}{548.4} = 1.01 \times 10^{-7}$$

Hence the ionic product of water.

$$K_w = [H^+][OH^-]$$

$$= 1.01 \times 10^{-7} \times 1.01 \times 10^{-7} = 1.02 \times 10^{-14}$$

(e) Degree of Dissociation of Weak Electrolyte : The degree of dissociation of ionisation α, of a weak electrolyte is given by

$$\alpha = \frac{\lambda_v}{\lambda_\infty} = \frac{\text{Equivalent conductance at the given dilution}}{\text{Equivalent conductance at infinite dilution}}$$

λ_r for a weak electrolyte is determined experimentally, while the λ_∞ is calculated with the help of Kohlrausch's law.

CONDUCTANCE IN NON-AQUEOUS SOLVENTS

Various substances (generally in liquid form) other than water which are used for dissolving the other compounds are known as *non-aqueous solvents*. In earlier studies of measurement of conductance, water was generally utilised due to:

(a) its wide liquid range

(b) ease of handling

(c) its plentiful occurrence in nature

(d) its unusual solvating characteristics and

(e) its high dielectric constant

The work on the measurements of conductances or substances in non-aqueous solvents first appeared in 1892.

Later on a large number of papers appeared in 1897 when Cady started his studies on liquid ammonia as a solvent. The conductance of non-aqueous solutions depend upon the following two factors:

(a) Effect of Dielectric Constant : The dielectric constant of a solvent can affect the dissociating power. This effect was realised by Thomson and Nernst and is known as Thomson-Nernst rule. According to this rule,

"The force exerted between the charged particles say e_1 and e_2 is inversely proportional to the dielectric constant (D) of the medium and to the square of the distance (d) between the particular."

$$F = \frac{1}{D}\frac{e_1 e_2}{d^2}$$

From the above relation it follows that E is the amount of energy required to separate two ions of charges e_1 and e_1 which are d cm apart. If the dielectric of the solvent is large the energy required to separate two ions will be smaller and therefore, the ionisation can take place easily.

Thus, formic acid, hydrogen cyanide, nitromethane, ethylene glycol, methyl alcohol, etc., with a high value of dielectric constant should behave as good ionising solvents whereas the solvents with a low dielectric constant such as hydrocarbons and halogenated hydrocarbons should be poorest ionising solvents.

In actual practice, we observe that the above expectations are in agreement with experimental observations which may be seen from the following Table 2.

Table 2

Solvent	*Dielectric Constant*	*Degree of Ionisation in 2000 litres dilution* $(C_2H_5)_3N$ *Picrate*	$(C_3H_5)_1NH$ *Picrate*	$(C_2H_5).NH_2.HCl$
H_2O	81.1	0.99	0.97	0.97
CH_3OH	35.4	0.94	...	0.92
C_2H_5OH	25.4	0.89	...	0.88
CH_3CN	36.4	0.95	0.90	0.31
$(CH_3)_2CO$	20.7	0.88	0.96	0.10
C_2H_5N	12.4	0.74	0.53	0.04
$CHCl_3$	10.9	0.47	0.005	0.00

In media of low dielectric constant, one observes, that for dilute solutions, the value of λ_r diminishes (in the usual manner) with increasing concentration of the solute. However, a minimum value of λ_r is attained and again there occurs an increase in λ_r as concentration increases. The dielectric constant of a solvent depends upon its dipole moment. Fajan has proposed that the ionisation is large in those solvents which contain a large cation and a small anion.

***(b)* Effect of Viscosity:** In 1906, Walden reported the quarternary ammonium salts like $(C_2H_5)_6NI$ can be dissolved in a number of solvents other than water. He also studied the conductance of solutions of those compounds and gave a relation connecting equivalent conductivity of the solution, viscosity and dielectric constant of the medium.

$$\lambda_\infty \eta = \text{constant} \quad ...(1)$$

where λ_∞ is the equivalent conductance of a solution in non-aqueous solvents and η is the viscosity of this solvent. Equation (1) is known as Walden's rule which may be stated in words as:

"*The product of limiting equivalent conductance of a given solute and viscosity of the pure solvent is constant.*"

The constant in equation (1) is independent of temperature and nature of the solvents and the value of $(C_2H_5)_4NI$ in more than 30 solvents at 0°C is 0.70, *i.e.*,

$$\lambda_{\infty}\eta = 0.70 \qquad ...(2)$$

The above rule of Walden does not apply to water and glycol as solvents. For water, $\lambda_{\infty}\eta = 1$ and the value of constant was observed to vary with the nature of the solute. From equation (1) it follows that one can calculate the equivalent conductance of a solute at infinite dilution from the viscosity of the solvent and the value of constant which can be determined.

Limitations of Walden's Rule

(1) In general, Walden's rule is not universally valid. It is not applicable to highly associated solvents and also not to the solvents of high viscosity.

(2) The value of constant in Walden's rule equation (1) is found to vary with variation of temperature.

Another Walden's Relation

Walden also showed that there is a relation among $(\lambda_{\infty} - \lambda_r)$, viscosity and dielectric constant. His mathematical relation is

$$\lambda_{\infty} - \lambda_r = \frac{K_e}{D\eta} \qquad ...(3)$$

where K_e is constant which depends only on the nature of the solute and its concentration but not on the nature of the solvent. He reported that the value of K_c is approximately given by

$$K_e \sim 51.5\ C^{0.45} \qquad ...(4)$$

where C is the concentration of the solute in gm. equivalent per litre.

Walden also reported that equation (1) may be modified for strong electrolytes in various solvents such as

$$= \frac{41.5}{D} C^{1/2} \qquad ...(5)$$

Equation (5) can be used successfully to interpret the conductivity of dilute solution of strong electrolytes in non-aqeuous solvents of relatively high dielectric constants.

For example, CH_3OH(D = 30.3): C_2H_5OH (D = 24.1);
$C_6H_5NO_2$ (D = 35.5);
NH_3 liquid at –40°C (D = 23).

Onsagar's Equation

This may be written in the form

$$\lambda_r = \lambda_\infty - (A + B\lambda_\infty\sqrt{C}) \qquad ...(6)$$

where A and B are constants dependent only on the nature of the solvent and the temperature. When the above equation was applied to non-aqueous solutions, it was observed to be in full agreement with theoretical requirements for the chlorides, and thiocynates of the alkali metals in methyl alcohol solution. Other electrolytes, such as nitrates, tetralkyl ammonium salts and salts of higher valency, however, exhibit appreciable deviations. These discrepancies become more pronounced for the solvents of law dielectric constants. In order to explain these deviations, equation (6) was modified to:

$$\lambda_r = \alpha[\lambda_\infty - (A + B\lambda_\infty)\sqrt{(\alpha C)}\,] \qquad ...(7)$$

Equation (7) may be written as

$$\lambda_r = \alpha\lambda_r' \qquad ...(8)$$

where

$$= \lambda_\infty - (A + B\lambda_\infty\sqrt{(\alpha C)}\,) \qquad ...(9)$$

From equation (9), it follows that is smaller than λ_r for solvents of low dielectric constants (non-aqueous solvents). From equation (8), it follows that for medium of low dielectric constant the degree of dissociation α is numerically equal to $\lambda_r/$ instead of λ_r/λ_∞ as proposed by Arrhenius.

Conductometric Titrations

Introduction: These can be applied to all titrations in which there occurs a sharp change in the conductivity at the end point. *The determination of the end point of a titration by means of conductivity measurement is termed as conductometric titration.*

In a conductometric titration the titrant is added from the burette and the conductivity readings corresponding to various increments of titrant are plotted against the volume of titrant. Two curves will be obtained which will intersect each other at a point, called "*end point*" or "*equivalence point*".

When the reaction is not quantitative, there is some curvature in the curve near the end point which may be due to hydrolysis, dissociation of the product or appreciable solubility in case of precipitation reactions.

It should be borne in mind that the resistance, and thus the conductance of an electrolytic solution is profoundly affected by changes in temperature. The resistance decreases by about 1% to 2% for each degree increase in temperature and for this reason it is desirable to carry out a conductometric titration at an approximately constant temperature. If absolute measurements are to be made a constant temperature bath is, of course, necessary.

The concentration of the titrant must be 10 times as the solution being titrated. This is done to keep the volume charge small. If it cannot be done, a correction to the reading must be applied, *i.e.*,

Actual conductivity = $\left(\frac{v + V}{V}\right)$ × observed conductivity where v is the volume of titrant or reagent added and V is the original volume.

ADVANTAGES OF CONDUCTOMETRIC TITRATIONS

Among the many advantages that conductometric titrations possess over the ordinary volumetric methods, we may mention the following:

(i) They are particularly useful in the case of coloured or turbid liquids, where ordinary indicators cannot function.

(ii) They can be employed for the analysis of dilute solutions and also for very weak acids.

(iii) No special precautions are necessary as the titration approaches the equivalence point, since the latter is found graphically.

(iv) It is not necessary to measure the actual conductance, as we may use any quantity that is proportional to it, *e.g.*, the reading on a ?Wheatstone bridge or resistance box. The same can be directly plotted against the volume of the alkali solution used up.

Apparatus for Conductometric Titrations

(a) *Measuring Circuits:* In order to prevent concentration changes due to reaction at the electrodes, the resistance of solution is generally measured with alternating currents and frequencies from 60 to 10000 c/sec have been used for ordinary conductance measurements, although, work at high frequencies finds may applications. Most of the circuits used are of the Wheatstone bridge type.

Stock has published an excellent critical review of the apparatus used for conductometric titrations, specially for microtitrations.

(b) *Commercial Apparatus* : Fairly simple commercially available equipment is accurate enough for most conductometric titrations. Its main advantages are that it is convenient and can be set up and used without any knowledge of electricity or electronics.

(c) *Electrodes and Cells* : A number of titration cells have been recommended of which those of Britton and Kano are the most practical. The simplest titration cell consists of a dipping electrode, a beaker of the appropriate size and a mechanical stirring device. For precipitation titrations, the electrodes are mounted vertically to prevent excessive amounts of precipitate from collecting n the electrode surfaces. Because a large effective surface area minimises polarisation, most conductivity electrodes are plated with a layer of platinum black. The platinisation of electrodes also tends to decreases the cell capacitance and allows the establishment of a sharp balance point. On the other hand, platinised electrodes have a greater tendency to adsorb substances from solution. Difficulties are encountered with dilute solution or solution containing surface active agents; bright platinum of lightly platinised electrodes are used in such cases.

(d) *Special Methods* : Although conductance is generally measured by means of a.c. apparatus, it has been shown by a number of workers that d.c. can be used provided that the cell has two unpolarisable electrodes, such as Ag/AgCl electrode. Eastmen has showed that direct current method is capable of yielding results that agree to 10% with those obtained with the usual a.c. method.

Differential Conductance Titration

Such titration have been investigated by Daval and Duval. They used two cells connected in two branches of a Wheatstone bridge, both cells contained identical sample solution and both were titrated with the same reagent, but one cell always had 0.05 to 0.1 ml more titranti in it than the other.

The titration curves for each cell were plotted on graph and their point of intersection gave one end point.

TYPES OF CONDUCTOMETRIC TITRATIONS

1. Acid-base Titrations

(a) *Conductometric titrations of strong acids and strong bases: A* typical strong acid/strong base titration may be exemplified by:

$$H^+ + Cl^- + Na^+ + OH^- \rightarrow Na^+ + Cl^- + H_2O.$$

The net result being the replacement of hydrogen ions by sodium ions. As we know that equivalent ionic conductance of hydrogenise is much greater than that of the sodium ion, so that the conductance of the solution at the beginning of titration is large but decreases up to the equivalence point. At the equivalence point, the conductance is equal to that of a pure solution of sodium chloride and is at a minimum. Beyond the equivalence point, the conductance increases in direct proportion to the excess of sodium hydroxide added.

The slope of the first branch of the titration curve, and thus sharpness with which the end point may be established, depends on differences between the equivalent ionic conductance of the hydrogen ion and the cation of the base used, in this case the sodium ion. As the equivalent ionic conductance of the hydrogen ion is much greater than that of any other cation, strong acids can almost always be successfully titrated to a conductometric end point.

On *special important advantage of* conductometric titration is that the *relative change in conductance is almost independent of the concentration of strong acid titrated.* Thus dilute solution can be titrated with almost the same accuracy as concentrated solution. The titration curve is displaced along the 1/R axis, but the sharpness is not generally affected. In this respect, conductometric acid-base titrations differ from potentiometric titrations.

Although the titration involving strong acids with strong bases was among the first conductometric titrations attempted, the conductometric method is not much used for such determinations, because a large number of indicator methods are available, even coloured solutions are now mostly titrated potentiometrically with a glass electrode.

(b) *Strong acid with a weak base:* Let us consider the titration of HCl (strong acid) with NH_4OH (weak base). When ammonium hydroxide is added to HCl, the conductivity decreases first due to the replacement of the fast moving H^+ ions by slow moving NH_4^+ ions.

$$H^+Cl^- + NH_4OH \rightarrow NH_4^+ + Cl- + H_2O.$$

After the end point is passed, further addition of NH_4OH does not change the conductance because NH_4OH, a weakly ionised electrolyte, has a very small conductivity compared with that of the acid or salt.

(c) *Weak acid with a strong base:* Let us consider the titration of the weak acid like CH_3COOH with strong base NaOH. When a small amount of NaOH is added to CH_3COOH, the conductivity decreases initially and then increases with the further addition of NaOH.

$$CH_3COO^- + H^+ + Na^+ + OH^- \rightarrow CH3COO^- + Na^+ + H_2O$$

When the neutralisation of acid is complete, further addition of alkali introduces excess of OH^- ions. The conductance of the solution, therefore, begins to increase more sharply.

(d) *Conductometric titration of weak acid and weak bases:* When weak acids are titrated with strong bases, curves such as 1 and 2. Before the equivalence point the conductivity is governed by the extent of ionisation (and by the concentration) the conductivity first decreases as the anion formed suppresses the ionisation, passes through a minimum, and then increases upon the equivalence point because of the conversion of non-conductivity weak acid into its conducting salt. Extensive information about the effect of the dissociation constant on the titration curves of both weak acids and bases has been compiled by Britton. Bruni and Eastman have carried out mathematical analysis of the titration curves of weak electrolytes.

Near the equivalence point, the graphs of such titrations are often curved because the incompleteness of reaction allows extra hydroxide ions (or hydrogen ions if a weak base is being titrated) to be present, thus increasing the conductance; the more incomplete the reaction, the more is curvature.

Conductometric titrations utilise measurements far enough removed from the end point for a fair amount of curvature near the end point to be tolerated. In fact many titration reaction that are too incomplete for potentiometric end point detection can be used conductometrically. For instance, boric acid can be titrated conductometrically with sodium hydroxide whereas the pH values change at end point is not large enough for potentiometric or indicator methods to be used. Weak acids with dissociation constants as small as 10^{-10} can readily be determined, especially in fairly concentrated solution.

End-point Calculation Method: After the conductometric titration curve has been plotted in usual way, estimate the end point by drawing two straight lines (the usual graphical method). Arbitrarily choose values for f_1, f_2, and f_3 such that f_1 and f_2 are before the end point and f_3 is after the end point and calculate f_4 by using the appropriate formula.

For a weak acid/weak base titration:

$$\frac{f_2 + f_1 - 2f_1 f_2}{(1 - f_2)(1 - f_1)} = \frac{f_4 + f_3 - 2}{(f_2 - 1)(f_4 - 1)}$$

For a strong acid/weak base titration,

$$\frac{1 - f_2 f_1}{(1 - f_2)(1 - f_1)} = \frac{f_4 + f_3 - 2}{(f_2 - 1)(f_4 - 1)}$$

Draw straight lines through f_1 and f_2 and through f_3 and f_4. The intersection of these lines is the new end point. Repeat the process using the new estimated end point. The new end points found in this way will converge rapidly so long as the first estimated end point was not very greatly in error.

2. *Replacement Titration*

When a strong acid reacts with the sodium or potassium salts of week acids, the weakest acid will be replaced first. Such a replacement reaction is exemplified by

$$HCl + CH_3COONa \rightarrow NaCl + C_3H_3COOH$$

The investigation of the replacement of weak acid from its salt was first investigated by Dutoit.

Analogous reaction for the replacement of the weak bases from their salts by strong bases can, of course, be cited.

Kolthoff has studied this type of reaction extensively, and curves for two of Kolthoff's system.. Curve I shows the increase in conductivity during the replacement of acetic acid by hydrochloric acid to form sodium chloride; the increase is due to the fact that the mobility of chloride ion is greater than that of acetate ion. After the equivalence point, the conductively increases more rapidly, owing to the additions of excess of hydrochloric acid. Kolthoff has calculated that accurate titrations with 0.01 M solutions are possible if the dissociation constant of the liberated acid is less than 5×10^{-5}; Lapshin has, however, titrated the salt of weak acids and bases whose dissociation constants are greater than 10^{-4}.

The main application of replacement titrations is in the determination of alkaloids. As alkaloids are weak bases that are often isolable in water, indicator methods do not work well.

3. Precipitation Titrations

Precipitation titrations can be effectively followed by conductometric methods, though not so well as acid-base titrations which are characterised by sharp breaks because both the hydrogen ion and hydroxyl ion have very high equivalent ionic conductances. In precipitation titrations, one pair of ions is substituted for another and as one experimenter has a choice of reagents, good results can usually be obtained. If a cation is to be precipitated, titrant whose cation has the smallest possible mobility is selected and, if an anion is to be precipitated, a titrant whose anion has as small a mobility as possible, in this way the maximum possible change in conductance during a titration is assured.

Let us consider the titration between $AgNO_3$ and KCl. The reaction involved may be represented as

$$K^+ + Cl^- + Ag^+ NO_3^- \rightarrow K^+ + NO_3^- + AgCl.$$

In the early stages of the titration , addition of silver nitrate, the conductance does not change very much because the Cl^- ions are replaced by NO_3^- ions; both have almost same ionic conductances. After the end point is passed, the excess of the added salt causes a sharp increase in conductance. If both the products of the reaction are sparingly soluble, the curve obtained takes up the form. An example of this is the reaction between $MgSO_4$ and $Ba(OH)_2$. The reaction is as follows:

$$Mg^{2+}SO_4^{2-} + Ba_2^{+} + 2OH^- ® Mg(OH)_2 + BaSO_4$$

In this type of titrations, conductivity decreases in the beginning but increases after the end point has passed.

Precipitate titrations cannot be performed very accurately. The main reasons for this are as follows:

(i) Due to slow separation of the precipitate.

(ii) Due to the removal of titrated solute by adsorption on precipitate.

Some efforts have been made to eliminate the errors that are due to adsorption, slowness of precipitation and electrode fouling, and are common to a number of conductometric precipitation titrations carrying out the titrations at elevated temperature, *e.g.*, in boiling solution, this apparently gives accurate results but there is insufficient evidence to judge these methods critically.

4. Complexometric Titrations

Conductivity has long been used in the study of complex formation. Werner, for example, supported many of the postulates of his coordination theory with conductivity measurements. Under controlled conditions conductivity measurements can be used to show the type of electrolyte, *e.g.* univalent, a complex salt gives in solution. Conductometric titrations have also been applied to the study of complex salts; for instance, Job titrated roseo-cobaltic sulphate $[Co(NH_3)_5H_2O]_2(SO_4)_3$ with barium hydroxide, finding two breaks in the titrations curve. Let us illustrate the complexometric titration by titrating KCl with $Hg(ClO_4)_2$. This curve shows two breaks one due to the formation of $HgCl_4^{2-}$ and the other one at the end of the reaction due to the formation of K_2HgCl_4.

$$Hg(ClO_4)_2 + 4KCl \rightarrow HgCl_4^{2-} + 2K^+ + 2KClO_4$$

$$HgCl_4^{2-} + 2K^+ \rightarrow K2HgCl_4.$$

6

IONIC MOBILITIES AND VELOCITIES

INTRODUCTION

Ionic mobility possesses a fixed value for each ion, no matter of which electrolyte it forms a part. It may be emphasised that ionic conductance is not the same thing as ionic mobility. These are, however, proportional to each other. If u_a^o and u_c^o are the actual velocities of positive and negative ions of a given electrolyte at infinite dilution, then

$$\lambda_a^o = ku_a^o \quad ...(1)$$

$$\lambda_c^o = ku_c^o \quad ...(2)$$

where k is a constant.

Absolute ionic velocity of an ion represents its *velocity in cms*$^{-1}$ *under a potential gradient of 1 volt cm*$^{-1}$. It has shown that under a potential gradient of 1 volt cm^{-1}, the value of constant k is the change in one g equivalent of the ion *i.e.* 96500 coulombs (or 1 Faraday). Mathematically,

$$\text{Absolute ionic mobility} = \frac{\text{Ionic conductance}}{96500} \quad ...(3)$$

Mobilities of H^+ and OH^-

These ions have exceptionally high ionic mobilities in hydroxylic solvents like water, alcohols etc. In water, H^+ is strongly hydrated forming H_3O^+, which is able to transfer H^+ to the neighbouring hydrogen bonded water molecules by rearrangement of hydrogen bonds. Thus, H^+ is transferred from one end of the series of the other end without actually migrating through the solution. Similarly, the high mobility of OH^- in water is due to H^+ transfer between OH^- and water.

Experimental Determination of Ionic Velocities

Ionic mobility can be measured directly by a technique based on the movement of a visible boundary between two electrolytes. Suppose it is required to determine the ionic velocity of K^+. The KCl may be one of the principal electrolyte and the indicator electrolyte may be taken as $CdCl_2$ having one (*i.e.*, Cl^-) ion common.

A decinormal solution of KCl is held over a solution of similar concentration of $CdCl_2$, in a tube of uniform bore. A current is passed through the solution. K^+ ions move towards the cathode. These are followed by the slow moving Cd^{2+} ions and the boundary moves. Let the boundary move through a distance, x cm in t second when the fall of potential in electric fields is E volts cm^{-1}. The mobility of H^+ is given by

$$u = \frac{x}{t \times E} \text{ cm s}^{-1}/\text{volt cm}^{-1} \qquad ...(4)$$

Ostwald's Dilution Law

As dissociation of weak electrolytes being a reversible process, the rate of formation of ions and their removal by recombination will be governed by the law of mass action. The applicability of the law of mass action to equilibria involving weak electrolytes was first shown by Wilhelm Ostwald and a form of it developed below in terms of dilution instead of concentration is known as Ostwald's dilution law which we shall discuss now. Consider the ion formation by an incomplete dissociation of an electrolyte AB.

$$\underset{C(1-\alpha)}{AB} \rightleftharpoons \underset{C\alpha}{A^+} + \underset{C\alpha}{B^-}$$

If C moles of the electrolyte (AB) are dissolved per litre of the solution and α is the degree of dissociation, then at equilibrium, the concentration of

A^+ ions $= C\alpha$

B^- ions $= C\alpha$

and AB molecules $= C(1 - \alpha)$.

Applying the law of mass action to the above equilibrium, we get

$$\frac{C\alpha \times C\alpha}{C(1-\alpha)} = K \text{ or } K = \frac{C\alpha^2}{(1-\alpha)} \qquad ...(1)$$

where K is called the dissociation constant. The choice of units for the value of K will depend on the number of ions formed from one dissociating molecule. The volume, V, of a solution containing one mole or one gram formula weight of an electrolyte is called the *dilution for the solution.* When a solution has C moles of an electrolyte per litre, dilution V of the solution is given by

$$V = 1/C \text{ litre per mole}$$

Equation (1) may now be written in terms of V instead of C values:

$$K = \frac{\alpha^2}{(1-\alpha)V} \text{ moles per litre.} \qquad (2)$$

Equation (2) is a *mathematical expression of Ostwald's dilution law* for equilibrium state. This law holds good for weak electrolytes since the value of 'K' is constant for them. Thus, Ostwald's dilution law is nothing but the law of mass action applied to ionic equilibrium existing in the solution of weak electrolyte. For weak electrolytes such as acetic acid or ammonium hydroxide, the value of the degree of dissociation is very small.

Therefore, $1 - \alpha \cong 1.$

The equation (1) then simplifies to $K = C\alpha^2$

$$\therefore \quad \alpha^2 = \frac{K}{C}$$

or

$$\alpha = \sqrt{(K/C)} \qquad ...(3)$$

From equation (3), it follows that the degree of dissociation of a weak electrolyte is inversely proportional to the square root of its concentration. For equation (2) when α is small,

$$\alpha = \sqrt{(KV)} \qquad ...(4)$$

From the above expression, it follows that the degree of dissociation of weak electrolyte is directly proportional to the square root of its dilution, (*i.e.*, the volume containing 1 gm mole of the electrolyte).

Experimental verification of the dilution law: The dilution formula may be experimentally verified by determining the value of K and substituting the value of α at different dilution V in the expression $\alpha^2/(1-\alpha)V$. The value comes out to be constant in case of weak electrolytes within reasonable limits.

Hence α, the degree of dissociation, is determined by substituting the value of equivalent conductivity at λ_r different dilutions V and the value of equivalent conductivity at infinite dilution λ_∞ in the formula:

$$\alpha = \frac{\lambda_v}{\lambda_\infty}$$

Determination of dissociation constant of acetic acid: Acetic acid is a weak electrolyte and when dissolved in water dissociates as

$$CH_3COOH \rightleftharpoons CH_3COO^- + H^+$$

If C gm moles of acetic acid are dissolved/litre of the solution and α is the degree of dissociation, then the dissociation constant

$$K = \frac{C\alpha^2}{(1-\alpha)}$$

If λ_V is the equivalent conductance of the CH_3COOH solution at the given dilution and λ_∞ that at infinite dilution, then

$$\alpha = \lambda_V/\lambda_\infty$$

The value of λ_V is determined by conductivity measurement at dilution V. The value of λ_∞ for acetic acid can be calculated with the help of Kohlrausch's law, *i.e.*,

$$\lambda_\infty \text{ for } CH_3COOH = \lambda \text{ for } CH_3COO^- + \lambda \text{ for } H^+$$

On substituting the values of λ_V, λ_∞ and V, the value of K can be calculated. At 20°C the value of K for acetic acid is found to be 1.8×10^{-5}.

Limitations of Ostwald's Dilution Law

(1) Ostwald's dilution law is obeyed by weak electrolytes only. Strong electrolytes like HCl, NaOH, salts etc. do not obey this law. In such cases the values of dissociation constant K are never constant but rapidly fall with dilution.

(2) The values of degree of dissociation of an electrolyte as determined from colligative properties and from conductance measurements often disagree.

(3) The law is a direct corollary from the Arrhenius theory. But the Arrhenius theory itself fails when applied to the strong electrolytes. The behaviour of the strong electrolytes, however, is explained better by the modern theory known as Debye—Huckely theory.

Basically, the failures of Ostwald's dilution law are due to inapplicability of Arrhenius theory of dissociation for some solutions of electrolytes.

DEGREE OF DISSOCIATION

The degree of dissociation *is the fraction fo the total number of molecules dissociated into ions, i.e.,*

$$\alpha = \frac{\text{No. of molecules dissociated into ions}}{\text{Total number of molecules taken}}$$

We know that the conductivity of a solution is due to the presence of ions in solution. Greater the number of ions, greater is the conductivity.

$\therefore$ λ_v, the equivalent conductivity at a particular dilution

$\propto$ No. of the ions present at this dilution

As ions are produced due to the dissociation of molecules of the electrolyte, $\lambda_v \propto$ No. of molecules dissociated into ions at a particular dilution.

Similarly, $\lambda_\infty \propto$ No. of ions at infinite dilution. ...(i)

As at infinite dilution all molecules are in ionic form

$\lambda_\infty \propto$ Total number of molecules taken. ...(ii)

Dividing (i) by (ii), we get

$$\frac{\lambda_v}{\lambda_\infty} = \frac{\text{No. of molecules dissociated } \textit{into ions at a particular dilution}}{\text{Total number of molecules taken}}$$

$$= \alpha \text{ (the degree of dissociation)}$$

Thus, *degree of dissociation, α, at any dilution, is the ratio of the equivalent conductivity at that dilution to the equivalent conductivity at infinite dilution.*

The value of λ_v can be obtained by direct measurement of conductivity; while λ_∞ can be calculated with the help of Kohlrausch's law, *i.e.*,

$$\lambda_\infty = \lambda^+ + \lambda^-$$

Hence, the value of α, the degree of dissociation, can be computed.

DISSOCIATION CONSTANT

According to Ostwald's dilution law for a weak binary electrolyte

$$\frac{\alpha^2}{(1-\alpha)V} = K$$

where, α is the degree of dissociation when 1 mole of electrolyte is dissolved in V litres of water. The constant K is called the dissociation constant or ionization constant of the electrolyte.

For the dissociation of acetic acid, a weak binary electrolyte, dissociation is represented as

$$\underset{(1-\alpha)\text{ moles}}{CH_3COOH} \rightleftharpoons \underset{\alpha\text{ g-ion}}{CH_3COO^-} + \underset{\alpha\text{ g-ion}}{H^+}$$

Suppose 1 g mole of acetic acid is dissolved in V litres of water. If α is the degree of dissociation at this dilution, then at equilibrium

$$[CH_3COOH] = (1-\alpha)V;\ [CH_3COOH^-] = [H^+] = \alpha/V.$$

Therefore, on applying law of mass action we get Oswald's dilution formula as

$$\frac{[CH_3COO^-][H^+]}{[CH_3COOH]} = \frac{\alpha^2}{(1-\alpha)\,V} = K_a$$

where K_a is the *dissociation constant* of acetic acid.

We know $\alpha = \lambda_v/\lambda_\infty$, therefore the above equation takes the form

$$\frac{\lambda_v^2}{\lambda_\infty(\lambda_\infty - \lambda_v)V} = K_a$$

The value of λ_v is determined by conductivity measurement at dilution V.

The value of λ_∞ for acetic acid can be calculated with the help of Kohlrausch's law, *i.e.*,

$$\lambda_\infty\ CH_3COOH = \lambda CH_3COO^- + \lambda H^+$$

One substituting the values of λ_y, λ_∞ and V, the value of K_a can be calculated. At 20°C the value of K_a for acetic acid is found to be 1.8×10^{-5}.

DEBYE-FALKENHAGEN EFFECT, CONDUCTANCE UNDER HIGH A.C. FREQUENCIES

Debye and Falkenhagen examined the conductance behaviour of a solution of a strong electrolyte by applying alternating currents of different frequencies. They predicted that *if the frequency of alternating current is high* so that *the time of oscillation is small* in comparison with the relaxation time of the ionic atmosphere, the asymmetric effect will be virtually absent. In other words, the ionic atmosphere around the central ion will *remain symmetrical*. The retarding effect due to asymmetry may, therefore, be entirely absent and the conductance may be higher. The conductance of a solution, therefore, should vary with the frequency of the alternating current used. The higher the frequency, the higher the conductance, evidently. This effect, also known as dispersion of conductance, has been verified experimentally. The conductance remains independent of the frequency of alternating current upto 10 cycles per second. But with further increase in frequency, the conductance starts increasing towards a certain limiting value indicating complete absence of asymmetric effect.

Wien Effect, Conductance under High Potential Gradients

Speed of an ion in an electric field varies with the applied potential gradient. Thus, under a potential gradient of about 20,000 volt per cm, an ion may have a speed of about 100 cm per sec. The ion, therefore, should *pass several times through the thickness of the ionic atmosphere during the time of relaxation.* The moving ion, therefore, will be almost free from the effect of the oppositely charged ionic atmosphere. The ion will be moving so fast that there will be no time for the ionic atmosphere to be built up. The asymmetry and electrophoretic effects, under these circumstances, may be negligibly small or even absent. Thus, the conductance of a strong electrolyte in aqeuous solution increases to a certain limiting value with increase in potential gradient applied. This observation had been verified experimentally by Wien much before the development of the theory of strong electrolytes and is known as the Wien effect.

STRENGTH OF AN ACID

According to the *ionic theory of ionization:*

(i) An acid is a substance which furnishes H^+ ions by itself or when H^+ dissolved in water.

(ii) The strength of an acid depends upon the concentration of free H^+ ions.

(iii) With the dilution, the degree of ionisation of an acid increases, thus increasing H^+ ions also.

(iv) An infinite dilution all acids are *completely ionised* and hence *they behave nearly equally strong at infinite dilution.*

(v) As degree of ionisation of an acid depends on its dissociation constant, so, the *dissociation constant of acid is the measure of its strength.*

(vi) Mineral acids like HCl, NHO_3, H_2SO_4, etc., which are largely ionised at all dilutions, are called *strong acids*, which oxalic, acetic, formic, benzoic acids, etc., which are *feebly ionised,* are called *weak acids*.

Methods of Comparing Relative Strengths of Two Acids

1. Thomson's Thermal Method

This method is based on the fact that when a mixture of two acids is allowed to react with a base, the quantity of heat evolved will be the measure of their relative strengths. Suppose we want to compare the strength of HCl and H_2SO_4.

Let the heat of neutralization of 1 g equivalent of HCl by NaOH

$$= x \text{ cal}$$

and heat of neutralization of 1 g equivalent of H_2SO_4 by NaOH

$$= y \text{ cal}$$

Now treat mixture containing 1 g equivalent of each HCl and H_2SO_4, with 1 g equivalent of NaOH and let the heat produced

$$= z \text{ cal}$$

The two acids will react with NaOH in the ratio of their relative strengths, the stronger acid taking more part in neutralizing it. Suppose n g equivalent of HCl and (1 – n) g equivalent of H_2SO_4 have neutralized 1 g equivalent of NaOH; then

Heat evolved due to neutralization of n g equivalent of HCl by NaOH+ heat evolved due to neutralization of (1 – n) g equivalent of H_2SO_4 by NaOH

$$= z \text{ cal}$$

or $$nx + (1 - n)y = z$$

or $$n = \frac{z - y}{x - y}$$

Knowing n, the relative strengths of $HCl/H_2SO_4 = n/(1 - n)$ can be calculated.

2. Conductivity Method

The strength of an acid at any dilution depends upon its degree of dissociation. So, the relative strengths of two acids at same dilution is equal to the ratio of their degrees of ionisation. For a weak acid, the degree of ionisation is given by conductivity ratio. Thus

For acid I: $\alpha_1 = \dfrac{\lambda_v}{\lambda_\infty}$

For acid II: $\alpha_2 = \dfrac{\lambda'_v}{\lambda'_\infty}$

But at infinite dilution

$$\lambda_\infty = \lambda'_\infty$$

$$\therefore \quad \frac{\alpha_1}{\alpha_2} = \frac{\lambda_v}{\lambda'_v} = \frac{\text{Strength of acid I}}{\text{Strength of acid II}}$$

Thus, *the relative strengths of two weak acids is equal to the ratio of their equivalent conductivities at the same dilution.*

3. Dissociation Constants Method

(For weak acids only).

For acids HA_1 and HA_2 at some dilution (V), we have

$$K_1 = \frac{\alpha_1^2}{(1-\alpha_1)V}; \quad K_2 = \frac{\alpha_2^2}{(1-\alpha_2)V}$$

For weak acids $(1 - \alpha_1)$ and $(1 - \alpha_2)$ may be taken as unity.

$$\therefore \quad \frac{\alpha_1^2}{V} = K_1; \quad \frac{\alpha_2^2}{V} = K_2$$

or

$$V = \frac{K_1}{\alpha_1^2} = \frac{K_2}{\alpha_2^2}$$

or

$$\frac{\alpha_1}{\alpha_2} = \sqrt{\left(\frac{K_1}{K_2}\right)} = \frac{\text{Strength of acid I}}{\text{Strength of acid II}}$$

Thus, *relative strength of two weak acids is equal to the square root of the ratio of their dissociation constants.*

4. Ostwald's Volume Method

In this method change in volume is noted instead of noting the heat evolved. Let x cm^3 be the change in volume when 1 equivalent of each of NaOH and HA_1 react and y cm^3 be the change in volume when 1 equivalent each of NaOH and HA_2 react. Let z cm^3 be the change when 1 equivalent of NaOH and a solution containing 1 equivalent of each HA_1 and HA_2 are mixed. Calculating in the same manner,

$$n = \frac{z - y}{x - y}$$

$$\therefore \quad \frac{\text{Strength of } HA_1}{\text{Strength of } HA_2} = \frac{n}{(1 - n)}$$

5. The Catalytic Method

It is well known that H^+ ions act as catalyst in many chemical reactions. The hydrolysis of esters and the inversion of cane sugar are catalysed by H^+ ions. To compare the strength of two acids HA_1 and HA_2, therefore, the velocity constants k_1 and k_2 of above first order reactions are determined in presence of these acids and the strengths are compared as

$$\frac{\text{Strength of } HA_1}{\text{Strength of } HA_2} = \frac{k_1}{k_2}$$

The strength of two bases can also be compared by the above mentioned methods. The conductivity method and comparison of dissociation constant method can be used as such. In catalytic process, we shall choose the base catalysed reactions such as condensation of acetone. In the Thomson's thermal and Ostwald's volume methods the two bases are neutralised by strong acids, say nitric acid.

DISSOCIATION CONSTANTS OF POLYBASIC ACIDS

Polybasic acids contain two or more ionisable hydrogen. They always dissociate in steps. The degrees of dissociation at different steps are different. Consider for example, the dissociation of phosphoric acid. It is a tribasic acid and ionises in three steps.

(i) $H_3PO_4 \rightleftharpoons H^+ + H_2PO_4^-$; $K_{a_1} = 7.5 \times 10^{-3}$ M

(ii) $H_2PO_4^- \rightleftharpoons H^+ + HPO_4^{2-}$; $K_{a_2} = 6.2 \times 10^{-7}$ M

(iii) $HPO_4^{2-} \rightleftharpoons H^+ + PO_4^{-3}$; $K_{a_3} = 4.8 \times 10^{-3}$ M.

The values of dissociation constants, K_{a_1}, K_{a_2} and K_{a_3} for the above three successive steps are taken from data book. These dissociation constants can be calculated in terms of the concentration, $[H^+]$, $[H_2PO_4^{1-}]$ and $[HPO_4^{2-}]$ and $[PO_4^{3-}]$ as follows:

$$K_{a_1} = \frac{[H^+][H_2PO_4^-]}{[H_3PO_4]}$$

$$K_{a_2} = \frac{[H^+][HPO_4^{2-}]}{[H_2PO_4^-]}$$

$$K_{a_3} = \frac{[H^+][HPO_4^{3-}]}{[HPO_4^{2-}]}$$

The ionisation constant in these three stages successively indicates that

$$K_{a_1} > K_{a_2} > K_{a_3}$$

The reason for the decrease in the dissociation constant values is that in the first dissociation the positively charged proton comes from a neutral molecule while in the second it is detached from a negatively charged molecule and in the third dissociation, it is detached from a doubly negatively charged molecule. The presence of negative charges makes the process difficult for the loss of proton.

COMMON ION EFFECT

Consider a simple weak electrolyte AB which ionises in solution as

$$AB \rightleftharpoons A^+ + B^- \quad ...(1)$$

On applying the law of mass action, we get

$$K = \frac{[A^+][B^-]}{[AB]} \quad ...(2)$$

Now, suppose to the above solution a strong electrolyte CB or AD *i.e.*, an electrolyte having A^+ or B^- as common ion with the electrolyte AB is added. These will ionise as

$$CB \rightleftharpoons C^+ + B^-$$

$$AD \rightleftharpoons A^+ + D^-$$

Suppose we have added AD to the equilibrium (1). Thus, the concentration of A in solution is increased. In order to keep K constant [equation (2)] at the given temperature, some of the A^+ added must combine with B^- to form unionised AB *i.e.*, the above equilibrium (i) is shifted backwards. In other words, the degree of ionisation of AB is suppressed.

This decrease in ionisation of a weak electrolyte by the addition of a strong electrolyte having an ion common with the weak electrolyte is known as common ion effect.

For example, addition of ammonium chloride suppresses the ionisation of ammonium hydroxide and hydrochloric acid suppresses the ionisation of hydrogen sulphide.

Applications of Common Ion Effect

(i) *In Analytical Chemistry*: In II group of qualitative analysis, the ionisation of weak acid H_2S

$$H_2S \rightleftharpoons 2H^+ + S^{2-}$$

is suppressed by the presence of strong acid HCl having a common ion H^+

$$HCl \overset{\longrightarrow}{\rightleftharpoons} H^+ + Cl^-$$

Due to the suppressed degree of dissociation of H_2S, the concentration of S^{2-} ions produced is low which causes the precipitation of sulphide of II group cations only.

Similarly, in III group, the ionization of weak base NH_4OH,

$$NH_4OH \rightleftharpoons NH_4^+ + OH^-$$

is suppressed by the prior addition of NH_4Cl which furnishes a high concentration of common ion NH_4^+.

$$NH_4Cl \overset{\longrightarrow}{\rightleftharpoons} NH_4^+ + Cl^-$$

Due to this low dissociation of NH_4OH, the OH^- concentration is quite low. This low OH^- ion concentration is sufficient only to precipitate hydroxides of Al^{3+}, Cl^{3+} and Fe^{3+}.

(ii) *Purification of Common Salt:* NaCl, as obtained from natural sources, is usually contaminated with small amounts of deliquescent impurities like $CaCl_2$ and $MgCl_2$. The purification of chloride solution is effected by passing HCl gas into a saturated solution of the impure sodium chloride solution is effected by passing HCl gas into a saturated solution of the impure sodium chloride.

The addition of HCl (containing Cl^-) causes the precipitation of only NaCl, but impurities remain in solution since the solution is not saturated so far as calcium and magnesium chlorides are concerned.

(iii) *Salting out of Soap:* In the soap manufacture, a solution of common salt is added to solution of soap (sod. stearate) to obtain solid soap. Addition of common salt (NaCl) increases the value of $[Na^+] \times [\text{stearate}^-]$ in solution until the solubility product of soap is exceeded, whereby soap gets precipitated.

(iv) *In Gravimetric Analysis:* When excess of precipitant (HCl or H_2SO_4) is added to solution containing Ag^+ or Ba^{2+} ions, complete precipitation of AgCl or $BaSO_4$ occurs due to the presence of excess of common ion Cl^- or SO_4^{2-}.

SOLUBILITY PRODUCT

Consider the *saturated* solution of a sparingly soluble binary electrolyte AB. There exist two equilibria simultaneously between undissolved solid, dissolved (but unionised) molecules and free ions, *i.e.*,

$$\underset{\text{(Undissolved solid)}}{AB} \overset{K_1}{\rightleftharpoons} \underset{\text{(Dissolved, but unionsed)}}{AB} \overset{K_2}{\rightleftharpoons} \underset{\text{(Ions in solution)}}{A^+ + B^-}$$

Applying the law of mass action, it follows that

$$\frac{[A^+][B^-]}{[(AB)_d]} = K_1; \quad \frac{[(AB)_d]}{[(AB)_d]} = K_2$$

∴ On multiplying, we get

$$\frac{[A^+][B^-][AB)_d]}{[(AB)_d][AB)_s]} = \frac{[A^+][B^-]}{[(AB)_s]} = K_1 \times K_3$$

But the concentration of solid AB is constant, so that

$$[A^+][B^-] = \text{a constant} = K_s$$

where the constant K_s is known as the solubility product of AB.

In general, for an electrolyte, A_xB_y, which ionises as

$$A_xB_y \rightleftharpoons xA^{y+} + yB^{x-}$$

the solubility product is given by

$$K_s = [A^{y+}]^x [B^{x-}]^y$$

∴ where $[A^{y+}]$ and $[B^{x-}]$ are the ionic concentration in a *saturated solution.* Thus, solubility product of an electrolyte may be defined as the *maximum product of the concentrations of its constituent ions (expressed in g-ion per litre) in its solution, when each ionic concentration term being raised to the number of times the ion occurs in the equation representing the solution of 1 molecule of the electrolyte.* For example for

$$Ag_2CrO_4 \rightleftharpoons 2Ag^+ + CrO_4^{2-}$$

the solubility product, $K_s = [Ag^+]^2[CrO_4^{2-}]$.

Relation Between Solubility and Solubility Product

Consider the general form of electrolyte A_xB_y which dissociates as

$$\underset{S}{A_xB_y} \rightleftharpoons \underset{xS}{x[X^{y+}]} + \underset{yS}{y[X^{x-}]}$$

Let S gram mole/litre be the solubility of the electrolyte, then

$$[A^{y+}] = x \,.\, S \quad \text{and} \quad [B^{x-}] = y \,.\, S$$

$$\therefore \quad K_s = [A^{y+}]^x[B^{x-}]^y$$

$$= [x \,.\, S]^x[y \,.\, S]^y = x^x \,.\, y^y \,.\, S^{x+y}$$

For example in case of $BaCl_2 \rightleftharpoons Ba^{2+} + 2Cl^-$; x = 1 and y = 2

$$Ks \rightleftharpoons 1^1 \,.\, 2^2\, S^{1+2} = 4S^3.$$

Analytical Applications of Solubility Product

From the relation:

$$K_s = [A^+]^x[B^-]^y$$

it is clear that:

(a) a solution in which $[A^+]^x[B^-]^y$ is *less than* K_s, the solution is unsaturated and more A_xB_y can be dissolved in it,

(b) a solution in which $[A^+]^x[B^-]^y$ is *equal* to K_s, *is just saturated*, and

(c) if anything is done to the solution which tends to make $[K^+]^x$ $[B^-]^y$ *greater than* K_s, then solid A_xB_y will be precipitated.

From the above discussion it is evident that if the product of concentration of ions in a saturated solution of an electrolyte becomes greater than the solubility product of the electrolyte, it will precipitate out. In qualitative analysis, separation and identification of metal ions such as Cu^{2+}, Zn^{2+}, etc., is possible only due to difference in solubility products of their sulphides.

The above concepts of solubility product have become extremely useful in getting a clear picture of different processes in analytical chemistry. A few cases are illustrated below:

1. Qualitative Analysis of Cationic Mixtures

The identification of individual species of cations present in an unknown mixture of electrolytes involves a preliminary segregation of the cations into selected classifiable groups.

A simple system devised for this purpose is based on the selective precipitation of sparingly soluble compounds of the cations from aqueous solution of the electrolyte mixture. In its most simplified form, for the more common cations only, these compounds fall into five groups. Within each group the anion is the same for the different cations and the solubility products are small and do not differ too greatly from one another:

Group I: The sparingly soluble chlorides derived from Pb^{2+}, Hg_4^{2+} and Ag^+ ions are precipitated by the addition of dilute HCl.

Chloride	K_s *at 25°C*
$PbCl_2$	1.6×10^{-5}
Hg_3Cl_2	1.3×10^{-15}
AgCl	1.8×10^{-16}

Group II: The sparingly soluble sulphides derived from Hg^{2+}, Pb^{2+}, Cu^{2+}, Cd^{2+} and Bi^{3+} ions are precipitated by the passage of H_2S through an acidic solution whose concentration of hydrogen ions is small.

Sulphide	K_s *at 25°C*
HgS	3×10^{-32}
PbS	8×10^{-28}
CuS	8×10^{-36}
CdS	7×10^{-27}
Bi_2S_3	1×10^{-96}

Group III: The sparingly soluble hydroxides derived from Fe^{2+}, Cr^{3+} and Al^{3+} ions are precipitated by the addition of ammonium hydroxide solution in the presence of excess ammonium chloride

Hydroxide	*K_s at 25°C*
$Fe(OH)_3$	6×10^{-28}
$Cr(OH)_3$	7×10^{-31}
$Al(OH)_3$	1.4×10^{-24}

Group IV: A further group of sparingly soluble sulphides, those derived from CO^{2+}, Ni^{2+}, Mn^{2+} and Zn^{2+} ions are precipitated by the passage of H_2S through a basic solution.

Sulphide	*K_s at 25°C*
CoS	8×10^{-23}
NiS	2×10^{-21}
MnS	1×10^{-11}
ZuS	8×10^{-25}

Group V: The sparingly soluble carbonates derived from Ca^{2+}, Sr^{2+} and Ba^{2+} ions are precipitated by the addition of ammonium carbonate solution in the presence of NH_4OH and NH_4Cl.

Carbonate	*Ks at 26°C*
$CaCO_3$	4.7×10^{-9}
$SrCO_3$	7×10^{-10}
$BaCO_3$	1.6×10^{-9}

The most common cations remaining in solution after the precipitation fo group V are Na^+, K^+ (all the simple salts of the alkali metals are water-soluble) and Mg^{2+} ions. The order of group precipitation is such that the precipitating reagents for later groups also precipitate in most cases cations in preceding groups, which must therefore the completely removed in the above prescribed order if they are not to interfere. For example,

(i) K_S of silver sulphide is 7×10^{-50} at 25°C. It means that silver sulphide will be precipitated in group II unless it has been completely removed as AgCl in group I.

(ii) Any of the cations normally precipitated as their sulphides in group II, will reappear in the group IV precipitate unless completely removed in group II.

(ii) K_S for $ZnCO_3$ at 25°C is 2.1×10^{-11}. Hence $ZnCO_3$ will be precipitated in group V unless it has been completely removed as ZnS in group IV.

2. The Separation of the Sulphides into Groups II and IV

The values of K_S for group II sulphide are smaller than those for group IV sulphides. Hence, it is possible to precipitate the sulphides of those cations which fall in group II selectively in the presence of those cations which fall in group IV if the $[S^{2-}]$ from the reagent H_2S is regulated such that it exceeds that required to reach the solubility products of the group II sulphides but is less than that required to reach the solubility produces of the group IV sulphides. This is done as follows:

H_2S is a weak diprotic acid for which:

(a) $$H_2S \rightleftharpoons H^+ + HS^-$$

$$\frac{[H^+][HS^-]}{[H_2S]} = K_{S_1} = 9.1 \times 10^{-3} \text{ at } 18°C$$

(b) $$HS^- \rightleftharpoons H^+ + S^{2-}$$

$$\frac{[H^+][S^{2-}]}{[HS^-]} = K_{S_2} = 1.2 \times 10^{-15} \text{ at } 18°C$$

Hence, on multiplying:

$$\frac{[H^+]^3[S^{2-}]}{[H_2S]} = 10^{-22}$$

$$\therefore \quad [S^{2-}] = \frac{[10^{-22}][H_2S]}{[H^+]^2}$$

If H_2S at 1 atmosphere is bubbled through water, for the saturated solution:

$$[H_2S] \sim 0.1 \text{ mole/litre}$$

$$\therefore \quad [S^2] = \frac{10^{-22} \times 10^{-1}}{[H^+]^2} = \frac{10^{-22}}{[H^+]^2}$$

i.e., the $[S^{2-}]$ is dependent on the $[H^+]$ and may be varied by varying the acidity of the solution.

If $\quad pH = 0(1\ MHCl)$

$\therefore \quad [H^+] = 1$ mole/litre

or $\quad [S^{2-}] = 10^{-22}$ mole/litre

If $\quad pH = 12(0.01\ M\ NaOH)$

$\therefore$ $[H^+] = 10^{-12}$ mole/litre

or $[S^{2-}] = 10$ mole/litre

Hence if the acidity is too low, the corresponding $[S^{2-}]$ is large enough to exceed the requirements of the solubility products of both the group II and the group IV sulphides which are precipitated together in group II.

Conversely, if the acidity is too high, the $[S^{2-}]$ ion is suppressed to a degree when it is lower than that required to reach the solubility products of even the group II sulphides. These, therefore, remain in solution and are precipitated in group IV where the $[S^{2-}]$ in the basic solution is too high.

In practice, if the $[H^+]$ of the solution, after the precipitation of the group I chlorides is adjusted to 0.3 M prior to passing H_2S, the group II sulphides are precipitated selectively.

(a) Selective precipitation of the group III hydroxide: Three hydroxides only are precipitated in group III, those of Fe^{2+}, Cr^{3+} and Al^{3+} whose K_s values lie between 10^{-30} to 10^{-33}. However, the K_s values for the hydroxides of those cations which are normally precipitated in group IV as their sulphide are also relatively small, though of a higher order than the above there:

Hydroxide	*K_s at 25°C*
$Co(OH)_2$	2×10^{-16}
$Ni(OH)_2$	2×10^{-15}
$Mn(OH)_2$	1.6×10^{-13}
$Zn(OH)_2$	7×10^{-18}

Now, for a solution fo NH_3 in water the following equilibria exist:

$$NH_3 + H_2O \rightleftharpoons NH_4OH \rightleftharpoons NH_4^+ + OH^-$$

for which $$\frac{[NH_4^+][OH^-]}{[NH_4OH]} = K_b = 1.8 \times 10^{-5} \text{ at } 25°C$$

$$\therefore \quad [NH_4^+][OH^-] = 1.8 \times 10^{-5} [NH_4OH]$$

But $$[NH_4^+] = [OH^-]$$

$$\therefore \quad [OH^-]^2 = 1.8 \times 10^{-5} [NH_4OH]$$

For 1 M NH_4OH :

$$[OH^-]^2 = 1.8 \times 10^{-5}$$

or $$[OH^-] = 4.24 \times 10^{-3} \text{ mole/litre}$$

For 0.1 M NH_4OH:

$$[OH^-]^2 = 1.8 \times 10^{-5} \times 10^{-1} = 1.8 \times 10^{-6}$$

or $$[OH^-] = 1.34 \times 10^{-2} \text{ mole/litre}$$

Hence, if a solution of NH_4OH alone is used to precipitate the hydroxides in group III, the $[OH^-]$ is high enough to exceed the requirements of the solubility products for the hydroxides of the cations which fall in group IV. These are therefore precipitated together with the normal group III hydroxides.

It is therefore necessary to suppress the $[OH^-]$ concentration to such an extent where it exceeds the requirements of the solubility products for the group III hydroxides but is lower than is necessary to reach the solubility products for the hydroxides of the group IV cations.

This is done by the addition of an excess of NH_4Cl, when a buffer solution is produced consisting of the weak base, NH_4OH, and a salt of the weak base and a strong acid, NH_4Cl. NH_4Cl, being a salt, is a strong electrolyte:

$$NH_4Cl \rightarrow NH_4^+ + Cl^-$$

The excess NH_4^+ ions depress the ionisation of NH_4OH in order that K_b for NH_4OH may be maintained. Hence the $[OH^-]$ is suppressed to a point where only the group III hydroxides are precipitated.

(b) Quantitative Analysis: The solubility products of the silver halides are small:

At 25°C, K_s of AgCl $= 1.8 \times 10^{-10}$

K_s of AgBr $= 5.2 \times 10^{-13}$

K_s of AgI $= 8.3 \times 10^{-17}$

Hence, if a solution fo the soluble $AgNO_3$ is added to a solution containing a soluble halide, the halide ions are quantitatively removed from solution as the sparingly soluble silver salt. The concentration of halide ions remaining in solution when precipitation is just complete is very low and may be neglected.

For example, if $AgNO_3$ solution is added to NaCl solution, AgCl is precipitated. When precipitation is just complete, for the ions remaining in solution:

$$[Ag^+][Cl^-] = K_s = 1.8 \times 10^{-10}$$

But $$[Ag^+] = [Cl^-]$$

$$\therefore \quad [Cl^-] = (1.8 \times 10^{-10})^{1/2} = (1.34 \times 10^{-5}) \text{ mole/litre}$$

Thus precipitation of sparingly soluble silver halides is made the basis of quantitative determinations of the concentrations of solution of soluble halides. A solution of the halides of unknown concentration is titrated with a standard solution of $AgNO_3$. The end point of the reaction occurs when precipitation fo the sparingly soluble silver halide is just complete.

3. Using Excess of Reagent

The use of excess of reagent is essential to ensure complete precipitation in gravimetric analysis. The insoluble substances like AgCl, $BaSO_4$, etc., have very slight solubility but by adding excess of precipitant their solubility product is further exceeded. Thus, the ionic product $[Ba^{2+}][SO_4^{2-}]$ becomes still higher than the solubility product of $BaSO_4$, by using large excess of H_2SO_4.

$$BaSO_4 \rightleftharpoons Ba^{2+} + SO_4^{2-}$$

$$H_2SO_4 \rightleftharpoons 2H^+ + SO_4^{3-}$$

4. Use of K_2CrO_4 as Indicator in $AgNO_3$ Chloride Titrations

In silver nitrate-chloride titration, the $AgNO_3$ is added from the burette into the chloride solution to which some K_2CrO_4 solution is added. At first the added $AgNO_3$ precipitates AgCl, but not Ag_2CrO_4 because AgCl has a lower solubility (0.00145 g per litre) than Ag_2CrO_4 (0.0285 g per litre). When sufficient $AgNO_3$ has been added to precipitate all the AgCl, subsequent addition causes the formation of brick-red Ag_2CrO_4 precipitate, which serves to note the end-point.

Limitations of Arrhenius Theory

Although the Arrhenius theory was quite successful in explaining the behaviour of wear electrolytes, yet it could not explain the various observed facts about the strong electrolytes some important limitations of this theory are as follows:

(i) Ostwald's dilution law which is based on Arrhenius theory fails completely where applied to strong electrolytes.

(ii) As strong electrolytes (KCl, NaOH, etc.) conduct electricity in molten states also, it means that their ionisation must have taken place even in the absence of water.

(iii) Arrhenius theory simply assumes the existence of ions in solution. It does not explain how and why ions are formed in the solution.

(iv) Arrhenius theory simply assumes the independent existence of ions in solution but does not take into account that the mobility of ions can be changed by electrostatic forces between oppositely charged ions lying side by side, particularly in solutions of moderate concentrations.

(v) It was observed that while there was good agreement between the values of α obtained by conductivity measurements and also from the study of colligative properties such as osmotic pressure, etc., in the case of univalent electrolytes like HCl, KCl, etc., but there were marked discrepancies in the value of bivalent electrolytes like $MgSO_4$, $CuSO_4$, etc.

(vi) The heat of neutralization of HCl and NaOH was found to be independent of concentration but their degree of dissociation as measured by conductivity measurements or study of colligative properties was found to decrease with increase in concentration. Thus, the heat of neutralisation should also decrease with increase in concentration.

(vii) Arrhenius theory could not explain the variation of transport number with concentration.

(viii) The values of equilibrium constant of strong electrolytes were found to vary with the concentration of the electrolytes.

THEORY OF STRONG ELECTROLYTES

In order to explain the above mentioned anomalies, the various theories were put forward from time to time but none of these could explain the behaviour of strong electrolytes. However, the accepted theory is that of Debye and Huckel (1923). This theory was later developed by Onsagar on the basis of interionic concept.

Debye-Huckel Theory

The failure of Oswald's dilution law in case of strong electrolytes has been explained satisfactorily both qualitatively and quantitatively by Debye-Huckel (1923). *This theory is based upon the following assumptions:*

(i) *All strong electrolytes are completely ionised in all dilutions i.e., their degree of ionisation is 100%.*

(ii) *The ratio* l_v/l_∞ *for strong electrolytes does not represent the degree of ionisation but it is only a conductivity ratio.*

(iii) *The increase in conductivity* (l_v) *of a strong electrolyte solution on dilution, is due to increase in the mobility of ions.*

(iv) *The two ions present in the solution obey the Coulomb's law, i.e.,*

$$F \propto \frac{Q_1 Q_3}{r^2}$$

where Q_1 *and* Q_2 *are the charges on the ions and r is the distance between them.*

(v) *Each ion is surrounded by an ionic atmosphere of oppositely charged ions. To this ionic atmosphere, solvent molecules are then attached. Thus, a positive ion is surrounded by negative ions and vice-versa. This ionic atmosphere may be assumed to be formed in the following manner.*

Let us consider a positive ion situated at A. Suppose there is small volume element dv at the end of a radius vector from the point 'A'. It is assumed that the magnitude of the distance 'r' is of the order of less than about 100 times as the diameter of the ion.

Due to the thermal movement of ions, sometimes there occurs an excess of positive ions and sometimes an excess of negative ions in the volume dv. If average time is taken, it will be found that there will be a negative charge density in the volume dv if a positive charge is assumed to be present at A. In this way every ion may be assumed to be present at A. In this way every ion may be assumed to have an ionic atmosphere of opposite sign. The net charge of the ionic atmosphere is equal in magnitude and opposite in sign to that of central ion.

Debye and Huckel took the interionic attraction into consideration and calculated the ratio of activity to the concentration of on ion in the dilution solution. This was done as follows:

Electrical potential at any point is defined as "*the work done in bringing a unit charge from infinity to the particular point.*"

If ψ is the potential at any given point in the vicinity of a positive ion, then the work done in bringing a positive

ion of valency Z_+ and carrying a charge ε to that point is given by $Z_+\varepsilon\psi$.

Similarly, the work done in bringing a negative ion is Z, $\varepsilon\psi$.

At an appreciable distance from the given ion, the value of ψ may be taken as zero. At this zero potential, if n_+^o and n_-^o are the concentrations of positive and negative ions per unit volume then by the Maxwell-Boltzmann's distribution law for particles in a field of varying potential, we can write

$$n_+ = n_+^0 e^{-(Z_+ \varepsilon\psi/KT)} \quad ...(1)$$

$$n_- = n_-^0 e^{-(Z_- \varepsilon\psi/KT)} \quad ...(2)$$

where n_+ and n_- are the concentrations of positive and negative ions at the given point under condition, K is the Boltzmann's constant.

It can be understood very easily that there are on an average more negative ions that positive ions in the vicinity of any positive ion and vice-versa. Thus it can be said that every ion is surrounded by an oppositely charged ionic atmosphere, *i.e.*, ions of opposite signs predominate in the ionic atmosphere.

The density of electricity ρ_e at any point may be taken equal to the excess of positive or negative electricity per unit volume at that point, *i.e.*,

$$\rho_e = (n_+Z_+\varepsilon) - (n_-Z_-\varepsilon)$$

$$\rho_e = \left[n_+^0 e^{-(Z_+\varepsilon\psi/KT)} Z_+\varepsilon\right] - \left[n_-^0 e^{-(-Z_-\varepsilon\psi/KT)} Z_-\varepsilon\right] \quad ...(3)$$

[From Eqs. (1) and (2)]

For a uni-univalent electrolyte $Z_+ = Z_- = 1$ and $n_+^0 + n_-^0 = n$, where n represents the number per c.c. of ions of either kind in the bulk of the solution. Thus, equation (3) becomes as:

$$\rho_e = n\varepsilon\left(e^{-\varepsilon\psi/KT} - e^{\varepsilon\psi/KT}\right) \quad ...(4)$$

If an assumption is made that $\varepsilon\psi/KT$ is small in comparison with unity, it is thus possible to write the experimental series and neglecting higher powers of $\varepsilon\psi/KT$ of equation (4). Therefore, equation (4) becomes as

$$\rho_e = -\left(\frac{\varepsilon^2\psi}{KT}\right)2n \qquad ...(5)$$

In the general case when Z_+ and Z_- are not necessarily unity, and if the solution may contain several kinds of ions, equation (5) may take the following form

$$\rho_e = -\frac{\varepsilon^2\psi}{KT}\Sigma n_1 Z_1^2 \qquad ...(6)$$

where n_1 and Z_1 represent the number per unit volume and valence of the ions of the *i*th kind. The summation is taken over for all the types of ions present in the solution and equation (6) is applicable irrespective of the number of different kinds of ions.

In order to solve Eq. (6) for ψ it essential to introduce Poisson's equation, a relationship between ρ_e and ψ, and this equation in rectangular coordinates is

$$\frac{\partial^2\psi}{\partial x^2} + \frac{\partial^2\psi}{\partial y^2} + \frac{\partial^2\psi}{\partial z^2} = -\frac{4\pi\rho_e}{D} \qquad ...(7)$$

Converting equation (7) into polar coordinates and putting $\partial\psi/\partial\theta$ and $\partial\psi/\partial\phi$ as zero, equation (7) becomes as

$$\frac{1}{r^2}\frac{\partial}{\partial r}\left(r^2\frac{\partial\psi}{\partial r}\right) = -\frac{4\pi\rho_e}{D} \qquad ...(8)$$

where D is the dielectric constant of the medium and $\partial\psi/\partial\theta$ and $\partial\psi/\partial\phi$ are zero because the distribution of potential about any point in the electrolyte must be spherically symmetrical, and consequently independent of the angles θ and ϕ.

Substituting the value of ρ_e from equation (6) in (8), we get

$$\frac{1}{r^2}\frac{\partial}{\partial r}\left(r^2\frac{\partial\psi}{\partial r}\right) = \frac{4\pi\varepsilon^2}{DKT}\psi\Sigma n_1 Z_1^2 \qquad ...(9)$$

or

$$\frac{1}{r^2}\frac{1}{\partial r}r^2\left(\frac{\partial\psi}{\partial r}\right) = k^2\psi \qquad ...(10)$$

where

$$k = \left(\frac{4\pi\varepsilon^2\Sigma n_1 Z_1^2}{DKT}\right)^{1/2} \qquad ...(11)$$

The general solution of equation (10) is given by

$$\psi = \left(\frac{A_e^{-kr}}{r} + \frac{B_e^{-kr}}{r}\right) \qquad ...(12)$$

where A and B are integration constants.

Calculation of A and B

As r increases, ψ decreases. At an infinite distance from the given ion, *i.e.*, r, ψ must approach zero *i.e.*, constants A and B must be zero. Hence, equation (12) consequently becomes as

$$\psi = \frac{Ae^{-kr}}{r} \quad ...(13)$$

The value of A can be calculated by the fact that when k = 0, the concentration becomes zero and, therefore, the potential is simply that of a single ion in the absence of any other charges, *i.e.*,

$$\psi = \frac{Z_1\varepsilon}{Dr} \quad ...(14)$$

Also k = 0; therefore equation (13) becomes as

$$\psi = \frac{A}{r} \quad ...(15)$$

Eliminating ψ between Eqs. (14) and (15), we obtain

$$\frac{Z_1\varepsilon}{Dr} = \frac{A}{r}$$

or
$$A = \frac{Z_1\varepsilon}{D} \quad ...(16)$$

Substituting the value of A in equation (13), we get

$$A = \frac{Z_1\varepsilon e^{-kr}}{Dr} \quad ...(17)$$

Expanding equation (17), we obtain

$$\psi = \frac{Z_i\varepsilon}{Dr}\left(1 - kr + \frac{k^2r^2}{2!} - ...\right) \quad ...(18)$$

As k is a function of the concentration, it means that the higher powers of kr may be neglected for very dilute solution. Hence, equation (14) becomes as:

$$\psi = \frac{Z_i\varepsilon}{Dr}(1 - kr)$$

$$= \frac{Z_i\varepsilon}{Dr} - \frac{Z_i\varepsilon k}{D} \quad ...(19)$$

From equation (19), it follows that

(i) The term $Z_i\varepsilon/Dr$ represents the potential at a distance r due to a single ion of charge $Z_i\ \varepsilon$ in medium of dielectric constant D, and (ii) the term $-Z_i\varepsilon k/D$ is then potential due to the other ions, *i.e.*, those forming the ionic atmosphere of the given ion. It is this extra potential which is related to the extra free energy of the ionic solution.

Physical Significance of k

The value of 1/k is regarded as the equivalent radius of the ionic atmosphere, *i.e.*, it possesses the dimensions of length and is generally termed as *Debye Length.* It is of the order of 10^{-8} cm of ordinary solutions. From equation (11), it follows, that the actual value of k depends upon the (i) the concentration of the solution, and (ii) the valances of the ions. For one molar aqueous solution of univalent electrolyte at 25°C, 1/k = 3.1 Å.

Energy of a charged body: It is given as:

$$\text{Energy of a charged body} = \frac{1}{2} \times \text{charge} \times \text{potential.}$$

Therefore, for an ion of charge (= $Z_i\varepsilon$), the energy possessed by virtue of its ionic atmosphere is given by

$$E_i = \frac{1}{2}(Z_i\varepsilon)\left(-\frac{Z_i\varepsilon k}{D}\right)$$

$$= \frac{1}{2}\frac{Z_i^2\varepsilon^2 k}{D} \qquad \text{...(20)}$$

where $(-Z_i\varepsilon k/D)$ is the potential due to the ionic atmosphere of the given ion [see equation (12)]. The corresponding energy for 1 gm. ion is obtained on multiplying by the Avogardo number N, so that

$$E_t = -\frac{NZ_i^2\varepsilon^2 k}{2D} \qquad \text{...(21)}$$

According to the definition of chemical potential, the chemical potential of a particular ion in an ideal solution is given by

$$\mu_i = \mu_i^0 + RT\log_e x_i \qquad \text{...(22)}$$

where x_i is its mole fraction in the given solution. For a non-ideal solution, one can write

$$\mu_i = \mu_i^0 + \log_e a_i$$

$$= \mu_i^0 + RT\log_e(x_i f_i) \qquad [\because\ a_i = x_i f_i]$$

$$= \mu_i^0 + RT \log_e x_i + RT \log_e f_i \quad ...(23)$$

where a_i is the activity and f_i is the activity coefficient.

The difference between Eqs. (23) and (22) is RT $\log_e f_i$ which is equal to the difference in the free energy change accompanying the addition or removal of 1 gm ion of the ionic species from a large volume of real and dilute solution. Thus, the difference of free energy, RT $\log_e f_i$, is regarded as equivalent to the electrical energy of the ion due to ionic atmosphere [Eq. (20)]. Hence equation (20) becomes as:

$$RT \log_e f_i = \frac{NZ_i^2 \varepsilon^2}{2D} \quad ...(24)$$

$$-\log_e f_i = +\frac{NZ_i^2 \varepsilon^2 k}{DRT} \quad ...(25)$$

Debye Huckel Limiting Law: From equation (11), we have

$$k = \left(\frac{4\pi\varepsilon^2 \Sigma n_i Z_i^2}{DRT} \right)^{1/2} \quad ...(26)$$

If in the above equation n_i is replaced by c_i N/1000 and R/N written for K, equation (26) becomes as:

$$k = \left(\frac{4\pi N^2 \varepsilon^2}{1000\ DRT} \Sigma c_i Z_i^2 \right)^{1/2}$$

$$= \left(\frac{8\pi N^2 \varepsilon^2}{1000\ DRT} \mu \right)^{1/2} \quad ...(27)$$

where $\mu = \frac{1}{2} \Sigma c_i Z_i^2$

Substituting equation (27) in (25), we get

$$-\log_e f_i = +\frac{N^2 \varepsilon^2}{R^{3/2}} \left(\frac{2\pi}{1000} \right)^{1/2} \frac{Z_i^2}{(DT)^{3/2}} \sqrt{(\mu)} \quad ...(28)$$

$$\text{or } -2.301 \log f_i = \frac{N^2 \varepsilon^2}{R^{3/2}} \left(\frac{2\pi}{1000} \right)^{1/2} \frac{Z_i^2}{(DT)^{3/2}} \sqrt{(\mu)} \quad ...(29)$$

$$\text{or} \quad \log f_i = -\frac{N^2 \varepsilon^2}{2.303\ R^{3/2}} \left(\frac{2\pi}{1000} \right)^{1/2} \frac{Z_i^2}{(DT)^{3/2}} \sqrt{(\mu)} \quad ...(30)$$

The values of universal constants N, ε, π and R as well as the numerical quantity substituted in eq. (30) to yield the following equation

$$\log f_1 = -1.823 \times 18^6 \frac{Z_i^2}{(DT)^{3/2}} \sqrt{(\mu)} \qquad ...(31)$$

For a given solvent and temperature, D and T have definite values which may be inserted; equation (31) then takes the general form

$$\log f_1 = -AZ_i^2 \sqrt{(\mu)} \qquad ...(32)$$

where A is constant for the solvent at the specified temperature.

Discussion: Equation (32) is known as Debye-Huckel limiting law. Some important points about this law are described below:

(i) This law expresses the variation of the activity coefficient of an ion with the ionic strength of the medium.

(ii) This law is called limiting law because it can only work if infinite dilution is approached.

(iii) From equation (32), it follows that activity coefficient of an ion should decrease with increasing ionic strength of the solution.

Limitation: Equation (32) cannot be tested in this form because there does not appear to be any experimental method of evaluating the activity coefficient of a single ionic species.

Mean Activity of an Electrolyte

Let us consider a binary electrolyte which dissociates into v ions, v^+ and v_-. Now the mean activity coefficient $f_{\neq}$ is related to individual ion activities by the following relation:

$$f_{\neq} = v\sqrt{(f_+ v + f_- r^-)}$$

or

$$\text{lof } f_{\neq} = \frac{v^+ \log f_+ + v^- \log f_-}{v^+ + v^-}$$

If Z_+ and Z_- are the valencies of cations and anions respectively then $v^+z^+ = v_-z_-$, then

$$\log f_{\neq} = \frac{Z_- \log f_- + Z_+ \log f_-}{Z_+ + Z_-}$$

Substituting the above value in eq. (32), we get

$$-\log f_{\neq} = AZ_+Z_- \sqrt{(\mu)} \qquad ...(33)$$

Equations (32) and (33) define a law which is known as Debye-Huckel Limiting law which is applicable to ideal solutions. According to this law,

"*The departure from ideal behaviour in a given solvent is equal to the ionic strength of the medium and the valencies of ions and is not dependent upon their chemical nature.*"

Tests of Debye-Huckel Theory

The Debye-Huckel theory may be used in the following ways:

(i) If $-\log f_{\pm}$ (33) is plotted against $\sqrt{(\mu)}$ for a dilute solution, straight lines should pass through origin with the slope of the curve AZ_+Z_- . This has been epxerimentally verified.

(ii) In 1921 Lewis and Randall gave an empirical rule, according to which the mean activity in dilute solution is same in solutions of same ionic strengths. It has been observed that the equation (33) is in agreement with the above empirical rule of Lewis and Randall.

(iii) Different workers determined activity coefficients for univalent electrolytes by various methods and found that activity coefficients are the function of square root of the concentration. For such substances, ionic strength is equal to the concentration. These results are obeying equation (33).

DEBYE-HUCKEL—ONSAGAR'S EQUATION

In the case of strong electrolytes the value of λ_v is much less than λ_∞. It was assumed to be due to the following reasons.

(i) *Relaxation Effect:* Due to inter-ionic forces each ion has a tendency to be surrounded by ions of opposite charge called the ionic-atmosphere. Let us suppose a negative ion be surrounded by the ionic atmosphere of positive ions. When an E.M.F. is applied, the negative ions move towards anode where by the ionic atmosphere of positive ions is left behind to disperse while a new ionic atmosphere is under formation in front of it. But the new ionic atmosphere is not formed at the same rate at which the old disperses and the latter takes more time called "*the relaxation time*".

It shows that in the rear of the moving ion there will be always an excess of ions of opposite sign. The ion, therefore, will always be dragged back. The effect thus decreases the mobility of the ion and is known as relaxation effect or asymmetric factor. Onsagar (1927) showed that the value of relaxation force may be given by

$$\text{Relaxation force} = \frac{\varepsilon^3 Z_i K}{6DkT} wV$$

where ε, Z, k and K have their usual significance while that of w and V are given below.

(ii) *Electrophoretic Effect:* The solvent molecules attached to ionic atmosphere move in direction opposite to that of central ion. Thus, they cause friction due to which the mobility of the central ion is retarded. This effect is called electrophoretic effect. On the basis of Stoke's law, Debye-Huckel calculated the following expression for the electrophoretic force on an ion of *i*th kind.

$$\text{Electrophoretic force} = \frac{KVK_i}{6\pi\eta} \cdot eZ_1$$

where η = viscosity coefficient of the medium.

K_i = coefficient of frictional resistance of the solvent opposing the motion of *i*th kind, and

On the basis of the above arguments Debye-Huckel and Onsagar derived the following expression, w is defined by

$$w = Z_+Z_- \frac{2q}{1+q^{1/2}} \qquad ...(34)$$

and the value of q is given by

$$q = \frac{Z_+Z_-(\lambda_+ + \lambda_-)}{(Z_+ + Z_-)(Z_+\lambda_- + Z_-\lambda_+)} \qquad ...(35)$$

It is now possible to equate to forces acting on an ion of the *i*th kind when it is moving through a solution with a steady velocity u_i; the deriving force due to applied electric field is $\varepsilon z_1 V$ and this is opposed by the frictional force of the solvent, $K_i U_i$, together with the electrophoretic and relaxation forces; hence

$$\varepsilon Z_i V = K_i U_i + \frac{\varepsilon z_1 K}{6\pi\eta} K_i V + \frac{\pi^3 Z_i K}{6DKT} wV \qquad ...(36)$$

Dividing throughout by $K_i V$ and rearranging, we get

$$\frac{U_i}{V} = \frac{\varepsilon Z_i}{K_i} - \frac{\varepsilon Z_i K}{6\pi\eta} - \frac{\varepsilon Z_i K}{6DKT} \cdot \frac{w}{K_4} \qquad ...(37)$$

If the potential gradient is taken as one volt per cm. *i.e.*, $V = \frac{1}{300}$, then

$$U_i = \frac{\varepsilon Z_i}{300\,K_i} - \frac{\varepsilon K}{300}\left(\frac{Z_i}{6\pi\eta} + \frac{\varepsilon^2 Z_i}{6DkT}\frac{w}{K_i}\right) \quad ...(38)$$

At infinite dilution K is zero and so equation becomes

$$U_i^o = \frac{\varepsilon Z_i}{300\,K_i}$$

But $\lambda_i = FU_i^o$

$$\therefore \quad \frac{\varepsilon Z_i}{300\,K_i} = \frac{\lambda_i^o}{F} \quad ...(39)$$

Again $U = \frac{\lambda_i}{\alpha F}$ ...(40)

Substituting equations (39) and (40) in (38), we get

$$\frac{\lambda_i}{\alpha F} = \frac{\lambda_i^o}{F} - \frac{\varepsilon K}{300}\left(\frac{Z_i}{6\pi\eta} + \frac{\varepsilon}{6DkT}\frac{\varepsilon Z_i}{K_i}w\right) \quad ...(41)$$

When the electrolyte is completely ionised, it means that $\alpha = 1$. Therefore, the above equation becomes as

$$\lambda_i = \lambda_i^o - \frac{\varepsilon K}{300}\left(\frac{Z_i}{6\pi\eta}F + \frac{300\varepsilon}{6DkT}\lambda_i^o w\right) \quad ...(42)$$

We also know $\frac{\varepsilon Z_i}{K_i} = \frac{300\lambda_i^o}{F}$. Therefore, the above expression reduces to

$$\lambda_i^o = \lambda_i^o - \left[\frac{29.15 Z_i}{(DT)^{1/2}\eta} + \frac{9.90\times10^5}{(DT)^{3/2}}\lambda_i^o w\right] \times \sqrt{(c_+ Z_+^2 + c_- Z_-^2)} \quad ...(43)$$

But $c = c_i z_i$, therefore

$$\lambda_i = \lambda_i^o - \left[\frac{29.15 Z_i}{(DT)^{1/2}\eta} + \frac{9.90\times10^5}{(DT)^{3/2}\eta}\cdot\lambda_i^o w\right] \times \sqrt{[c(Z_+ + Z_-)]} \quad ...(44)$$

We know, equivalent conductance of the electrolyte is equal to the sum of conductances of the constituent ions and so it follows from equation (44) that

$$\Lambda = \Lambda_0 - \left[\frac{29.15\,(Z_+ + Z_-)}{(DT)^{1/2}\,\eta} + \frac{9.90 \times 10^5}{(DT)^{3/2}}\Lambda_o w\right]\sqrt{[c(Z_+ + Z_-)]} \quad ...(45)$$

In uni-univalent electrolyte, $Z_+ = Z_- = 1$ and $w = 2 - \sqrt{2}$, equation (45) becomes as

$$\Lambda = \Lambda_0 = \frac{82.4}{(DT)^{1/2}\,\eta} + \frac{8.20 \times 10^5}{(DT)^{3/2}}\Lambda_0\Big]\sqrt{c} \quad ...(46)$$

For a uni-univalent electrolyte, equation (46) becomes as

$$\Lambda = \Lambda_0 - [A + B_{\Lambda 0}]\sqrt{c} \quad ...(47)$$

where A and B are constants dependent upon the nature of solvent and the temperature, thus

$$A = \frac{82.4}{(DT)^{1/2}\,\eta} \quad ...(48)$$

$$B = \frac{8.20 \times 10^5}{(DT)^{3/2}} \quad ...(49)$$

Testing of Onsagar's Equation (49) : When equivalent conductance is plotted against the square root of concentration, a straight line should be obtained with slope $A = B\ \lambda_0$. These values closely agree with experimental data. Onsagar equation is obeyed at the concentration of about 2×10^{-3} equivalent per litre. This equation is also applicable to non-aqueous solvents.

Applications of Debye-Huckel Equation: The various forms of the equations resulting from the Debye-Huckel theory find practical application in the determination of activity coefficients and make possible the determination of thermodynamic data. Two important cases will be considered here.

1. *Determination fo thermodynamic equilibrium constants:* Let us consider as an example the dissociation of a 1 : 1 weak electrolyte AB

$$AB \rightleftharpoons A^+ + B^-$$

The thermodynamic dissociation constant k_T is given by

$$k_T = \frac{[A^+][B^-]}{[AB]}\frac{\gamma_A^+ \gamma_B^-}{\gamma_{AB}}$$

$$k_T = k\frac{\gamma \pm^2}{\gamma AB}$$

$$k_T \sim k\gamma\pm^2 \quad ...(1)$$

where k is the concentration or conditional dissociation constant. For a weak electrolyte in dilute solution γ_{AB} for the undissociated, and therefore non-ionic species is very nearly unity. Taking logarithms of Equation (1) and substituting for $\gamma\pm$ form the limiting law expression, we obtain

$$\log k = \log k_T + 2A\sqrt{\mu} \quad ...(2)$$

k_T may therefore be determined from measured values of over a range of ionic strength values and extrapolating the k versus $\sqrt{\mu}$ plot to $\sqrt{\mu}$ = 0. This is a general technique for the determination of all types of thermodynamic equilibrium constants, *e.g.*, solubility, stability and acid dissociation constants.

2. *Effect of ionic strength on ion reaction rates in solution:* In the treatment of ionic reactions by Bronsted and Bjerrum an equilibrium is considered to exit between reactant ions and a critical complex, the latter bearing close resemblance to the activated complex of the theory of absolute reaction rates. Thus for the reaction scheme,

$$A_z^A + B_z^B \rightleftharpoons X_z^{(A} + {}_z^{B)} \pm \rightarrow \text{Products} \quad ...(3)$$

We may write, for the per-equilibrium

$$k = \frac{[X\pm]}{[A][B]}\frac{\gamma \pm}{\gamma_A \gamma_B} \quad ...(4)$$

Omitting charges for clarity so that the rate, v with which A and B react may be expressed by

$$v = k[A][B] = k_0[A][B]\frac{\gamma_A \gamma_B}{\gamma \pm} \quad ...(5)$$

where $$k = k_0\frac{\gamma_A \gamma_B}{\gamma \pm} \quad ...(6)$$

k_0 being the specific rate in infinitely dilution

where $$\frac{\gamma_A \gamma_B}{\gamma \pm} = 1 \quad ...(7)$$

In logarithmic form equation (7) becomes

$$\log k = \log k_0 + \log \gamma_A + \log \gamma_B - \log \gamma_\pm \quad ...(8)$$

in which activity coefficients may be expressed by the following equation,

$$\log k = \log k_0 - \frac{A\sqrt{\mu}}{1 + Ba\mu}[Z_A^2 + Z_B^2 - (Z_A + Z_B)^2] \quad ...(9)$$

$$= \log k_0 + \frac{2AZ_AZ_B\sqrt{\mu}}{1 + Ba\sqrt{\mu}} \quad ...(10)$$

$$\log k \sim \log k_0 + 2AZ_AZ_B\sqrt{\mu} \quad ...(11)$$

in very dilute solution,

or

$$\log k \sim \log k_0 + 1.02\ Z_AZ_B\sqrt{\mu} \quad ...(12)$$

for water as solvent at 298 k.

These last equations take account of the salt effect observed for reactions between ions, the slopes of graphs of log k/k_0 versus $\sqrt{\mu}$ being very close to those predicted by Equation (3) at low concentration. At higher concentrations, deviations from linearity occur and these are particularly noticeable for reactions having $Z_AZ_B = 0$, *e.g.*, for a reaction between an ion and a neutral molecule. According to Eq. (12), such reactions should show no variation of rate with ionic strength and this is indeed the case until about $\mu = 0.1$. Above this point, increasing ionic strength does cause the rate to vary.

IONIC PRODUCT OF WATER

Pure water is essentially a covalent compound. Nevertheless, it ionises very slightly and the following equilibrium is established:

$$H_2O \rightleftharpoons H^+ + OH^- \quad ...(1)$$

where H^+ is a hydrogen ion and OH^- is a hydroxyl ion. The removal fo the one orbital electron from a hydrogen atom yields the positive hydrogen ion or hydrogen cation, which is in fact a bare proton. The bare proton has no separate existence in solution and owes its stability to solvation by a water molecule to give the hydronium ion, H_3O^+. The ionisation equilibrium of water is thus represented more accurately as:

$$2H_2O \rightleftharpoons H_3O^+ + OH^-$$

However, as many qualitative and quantitative considerations remain essentially the same for the unsolvated and the solvated proton, it is usual for convenience to refer to the unsolvated proton, H^+.

On applying the law of mass action to equation (1), we get

$$K = \frac{[H^+][OH^-]}{H_2O} \qquad ...(2)$$

where, K is called the ionisation constant. As water is ionised to very slight extent (about 1 in 10^7 water molecules), the concentration of water molecules is very large as compared with H^+ and OH^- ions so that it can be regarded as practically constant.

Therefore, equation (2) may be written as

$$[H^+][OH^-] = K[H_2O] = k_w \qquad ...(3)$$

where constant k_w is constant called the *ionic product of water.*

Conductivity measurements show that in pure water,

$$[H^+] = 10^{-7} \text{ mole/litre at } 25°C$$

Furthermore in pure water,

$$[H^+] = [OH^-]$$

Therefore, equation (3) may become as

$$[10^{-7}][10^{-7}] = k_w$$

or $$k_w = 10–14 \text{ at } 25°C \qquad ...(4)$$

Significance of Equation (3) and (4)

For all aqueous solutions, the product of $[H^+][OH^-]$ may be taken to be equal to 10^{-14}.

If extra H^+ ions are introduced, the ionisation of H_2O is suppressed to the point at which $[H^+][OH^-] = 10^{-14}$. Similarly, if extra OH^- ions are introduced, the ionisation of water is again suppressed to maintain the constancy of $[H^+][OH^-]$.

From equation (3), it follows that the ionic product of water may be defined as *the product of concentration of H^+ and OH^- ions expressed in gram ions per litre.*

pH VALUE

From equation (3) and (4), we have

$$[H^+][OH^-] = 10^{-14} \qquad ...(5)$$

In case of pure water, $[H^+] = [OH^-]$

Therefore, equation (5) becomes as

$$[H^+][H^+] = 10^{-14}$$

or $$[H^+] = 10^{-7}$$

In means if the concentration of H^+ ions in solution is 1×10^{-7} gm, ion per litre, it is natural and if more, it is acidic and if less it is alkaline. Thus the acidity or alkalinity of the solution can be expressed in terms of its H^+ ion concentration.

It is not very convenient to express acidity and basicity in terms of H^+ ions as its value are usually very small, especially in case of weakly ionised substances. In order to overcome this difficulty, *Sorensen*, a Danish biochemist in 1909 introduced a new notation to express the H^+ ion concentrations. According to his the pH of a solution is,

"*numerically equal to the negative power to which 10 must be raise din order to express the H^- ion concentration*".

$$[H^+] = 10 \quad \text{or} \quad \log[H^+] = -\text{pH} \log 10$$

or $$\log [H^+] = -\text{pH} \qquad [\because \log 10 = 1]$$

or $$\text{pH} = -\log [H^+]$$

Hence pH is the *negative logarithm of hydrogen ion concentration.* For pure water or a neutral solution in which

$$[H^+] = 1 \times 10^{-7}$$

$$\text{pH} = -\log[H^+] = -\log[10^{-7}] = 7$$

The pH of neutral solution is 7. In acidic solution $[H^+] > 10^{-7}$ and therefore $\text{pH} < 7$. In alkaline solution $[OH^-] > [H^+]$ and therefore $[OH^-] > 10^{-7}$ and $[H^+] < 10^{-7}$ and, therefore, solution is acidic, and if it is more than 7, the solution is alkaline. A pH scale ranges from 0 to 14.

Determination of pH Value

(1) *Colorimetric Method (For Colourless Solutions Only):* If the same amount of the same indicator gives the same colour in the same amount of two different solutions, then the pH of the two solutions must be the same. The colorimetric method (or indicator method) is the simplest method requiring least expensive equipments and is applied where both speed and moderate accuracy are expected.

The method consists in adding a definite quantity of *universal indicator* (a mixture of certain dyes which give different colours at different pH) to measured quantity of solution under test contained in a glass tube. The colour so developed is compared with a series of colours obtained by adding same amount of indicator to standard buffer solutions of known but varying pH values. If a complete match is obtained, the pH of the unknown solution will be the same as that of the standard solution. The matching of colours can be best done by using a colorimeter. The use of a colorimeter with a set of coloured glasses provides a quick and simple method for measuring pH values.

(2) Emf Method: This is the most accurate method for determining the pH value (or H^+ ion concentration). The solution whose pH is to be determined is taken in a vessel of suitable shape. In it is dipped a plantinised platinum electrode and a steady stream of moist hydrogen gas is passed through the solution. This half cell so formed is connected to a saturated calomel electrode through a salt bridge and the emf of the complete cell is determined by means of a potentiometer. The pH value is then calculated using the formula

$$\text{pH} = \frac{\text{Emf} - 0.2422}{0.0591}$$

Indicators

Indicators *are organic substances the presence of very small amount of which indicates the termination of a chemical reaction by a change of colour.* Indicators are of various types, *e.g.*, acid-base indicators, redox indicators, adsorption indicators, etc. *Acid-base indicators* are organic substances which have one colour in acid solution while an altogether different colour in alkaline solution. Various acid-base indicators show colour changes in a definite pH range, *e.g.*,

Table 1

Indicator	*pH range*	*Colour in acid*	*Colour in alkali*
Photophthalein	8.3—10.5	Colourless	Red
Litmus	5.5—7.4	Red	Blue
Methyl red	4.5—6.5	Red	Yellow
Methyl orange	3.2—4.5	Pink	Yellow

THEORIES OF INDICATORS

1. Ostwald's Theory

According to this theory:

(1) Acid-base indicators are *weak organic acids or bases.*

(2) *They possess different colours in ionised and un-ionised states i.e.,*

$$\underset{\text{(one colour)}}{HIn} \rightleftharpoons H^+ + \underset{\text{(different colour)}}{In^-}$$

(3) *The colour of the indicator depends on the relative proportions of the unionised indicator molecules and its ions.*

Thus, *phenolphthalein* is a weak acid whose unionised molecules are colourless, while ions are red in colour *i.e.*

$$\underset{\text{(colourless)}}{HPh} \rightleftharpoons \underset{\text{(red)}}{Ph^-} + \underset{\text{(colourless)}}{H^+}$$

In presence of an acid (*i.e.*, H^+ ions) the equilibrium is forced backward (due to common H^+ ion), whereby resulting in the formation of colourless undissociated molecules. Addition of a strong alkali (*i.e.*, OH^- ions) results in the removal of H^+ ions.

$$\underset{\text{(of indicators)}}{H^+} + \underset{\text{(of added alkali)}}{OH^-} \rightleftharpoons \underset{\text{(nearly unionised)}}{H_2O}$$

This results in more and more dissociation of indicator. Therefore, in presence of an alkali the concentration of pink coloured Ph^- ions is far greater than colourless HPh molecules, and hence the solution will be pink.

If a weak base (*e.g.*, NH_4OH) is added, the OH^- ions furnished by it are very small in number, and hence the equilibrium is not shifted sufficiently to produce a large number of coloured Ph^- ions, *i.e.*,

$$HPh \rightleftharpoons H^+ + Ph^-$$

$$\underset{\text{(feebly ionised)}}{NH_4OH} \rightleftharpoons OH^- + NH_4^+$$

$$OH^- \downarrow H_2O \qquad NH_4^+ \rightleftharpoons \underset{\text{(Readily ionised)}}{NH_4Ph}$$

Moreover, NH_4Ph molecules formed are highly ionised to yield back NH_4^+ ions, which further reduces the number of Ph^- ions. Therefore, pink colour does not appear until a large excess of weak base is added to get visible colour change (or end-point).

Hence, this explain why *phenolphthalein is not a good indicator when weak base is used.*

Methyl orange is a weak base which ionises to yield red ions, *i.e.*,

$$\underset{\text{(Yellow)}}{MeOH} \rightleftharpoons \underset{\text{(red)}}{Me^+} + \underset{\text{(colourless)}}{OH^-}$$

If a base (*i.e.*, OH^- ions) is added to the indicator, the OH^- ions will suppress the ionization of the indicator. Hence, the indicator will remain yellow in an alkali. However, if a small excess of acid (say, HCl) is added, the latter will force the equilibrium to the right by removing OH^- ions to form H_2O.

$$\underset{\text{(of indicator)}}{OH^-} + \underset{\text{(of added acid)}}{H^+} \rightleftharpoons \underset{\text{(practically unionised)}}{H_2O}$$

This will result in the formation of red coloured Me^+ ions in the solution.

It may be pointed that methyl orange is not a suitable indicator for titrating a *strong base against weak acid* like CH_3COOH. After the end point, the weak acid does not produce sufficient H^+ ions to shift the equilibrium appreciably to the right. Moreover, the slat formed (CH_3COOMe) although it is highly ionised, yet it does not produce sufficient Me^+ ions due to hydrolysis.

$$CH_3COOMe + H_2O \rightleftharpoons CH_3COOH + MeOH$$

Thus, the colour does not appear until sufficient excess of CH_3COOH has been added. Consequently the end point will not be correct.

2. *Modern Quinoid Theory*

According to it:

1. *An acid-base indicator is a dynamic equilibrium mixture of two alternative tautometric forms*; ordinarily one form in *benzenoid* while the other is *quinoid.*
2. *The two forms have differents colours.*
3. *Out of these one form exists in acidic solution, while the other in alkaline solution.*
4. *Change in pH causes the transition of benzenoid form to quinoid form and vice-versa* and consequently a change in colour. The colour changes in case of methyl orange and phenolphthalein.

Indicators and acid-base titrations: A process of acid-base titration is accompanied by a change in pH when successive amounts of base are added to a solution of an acid (or *vice-versa*). Indicators have the property of changing colour in dilute solutions when hydrogen ion concentration of the solution attains a definite value. The colour charge in the indicator occurs within a certain pH range. Let us consider an indicator such as HIn, which is a weak acid and its ionisation is represented as

$$\underset{\text{Colour A}}{HIn} \rightleftharpoons H^+ + \underset{\text{Colour B}}{In^-}$$

Application of law of mass action to this reversible reaction gives

$$\frac{[H^+][In^-]}{[HIn]} = K_1$$

The colour exhibited by the indicator is determined by the ratio of the concentrations of the two species HIn and In^-.

Thus, $$\frac{[In^-]}{[HIn]} = \frac{K_1}{[H^+]}$$

$$\log K_1 - \log[H^+] = \log\frac{[In^-]}{[HIn]}$$

$$pK_1 + pH = \log\frac{[In^-]}{[HIn]}$$

(i) When pH = pK_1, the ratio $\frac{[In^-]}{[HIn]}$ becomes equal to 1 *i.e.* 0.5 indicator is present in the acid form and 0.5 in the alkaline form.

(ii) When pH = $pK_1 - 1$, then [HIn] will be ten times and at pH = $pK_1 - 2$, 100 times $[In^-]$. At these pH values the colour of the undissociated indicator will predominate.

In general the useful range of pH is $pK_1 \pm 0.5$ to 1 unit.

Titration Curves

A plot between pH of the solution during titration and the amount of acid/base added is called titration curve. Plots of pH vs. volume of acid or base added in an acid-base titration are useful in that they show the equivalence point graphically and help in the choice of a proper indicator.

Titration curves for various acid-base pairs of varying strength: The nature of the titration curve depends on the ionization constants of the acid and base employed in a titration, *i.e.*, on the strength of the acids and bases. We shall discuss now the nature of titration curves of various acid-base parts of varying strength.

(i) *Titration of a strong acid against a strong base:* The pH-titration curve of a solution of a strong acid (say HCl) against a solution fo a strong base (say NaOH). In the beginning when alkali is introduced into the acid solution, there is not much change in the pH value of the solution. This is in conformity with our earlier experience that strong acid solutions are good buffers at low pH. So the curve is almost flat until the end point is reached by successive additions of alkali when the sudden rise in pH of the solution takes place. The curve at this stage is almost vertical and parallel to pH axis.

The end point is also called the inflection point. On adding alkali after the inflection point, the sudden rise continues and soon the curve again becomes flat. This method can therefore, be used for determining the end point in case of the titrations involving solutions whose pH can be found out directly. In such titrations the pH is determined after successive additions of alkali and a graph is drawn between pH and the amount of alkali (volume of alkali) added to the acid solution. The inflection point (the point of the rapid rise in the curve) is the equivalence mid-point for the titration and represents the stage of titration where equivalent quantities of acid and base are present in the titration mixture. For the titration of a strong acid and a strong base, the equivalence point occurs at a pH 7.

In the titration of a strong base against strong acid the titration curve begins at high pH and remains flat till the end point is reached when there is a sudden fall in pH. Further addition of acid does not bring about an appreciable change in the pH of the solution and the curve is again flat. Here again the inflection point of the curve gives the end point.

(ii) *Titration of a weak acid against a strong base:* The pH titration curve of solution of a weak acid (say acetic acid) against the solution of a strong base (say NaOH). Before adding alkali to the acid solution, its pH can be calculated from its ionization constant and its concentrations On adding alkali in the beginning, the free hydrogenion concentration decreases due to the neutralisation by hydroxyl ions furnished in the solution by strong base and there is a big change in pH.

End point in this case also is given by the inflection point of the curve but the jump in pH in this case is smaller than that in the case of pH-titration curve when a strong acid and strong base are involved in the titration. The jump in pH depends on the pK_a of the acid. Higher the pK_a of the acid, lower is the jump in pH at the end point it will be seen that vertical portion now begins beyond pH 7 and end point lies between pH

8 and 10. This is due to the hydrolysis of the salt which gives OH^- ions in aqueous solution. In case of the titration of a solution fo strong base against a solution fo a weak acid, the curve starts with a higher pH and ends at a lower pH and the inflection gives the end point.

The nature of this curve is the same otherwise.

(iii) *Titration of a strong acid against a weak base:* There is not much difference in the titration curve of a strong acid against a weak base from the titration curve of a strong acid against a strong base till the end point is reached.

Here the vertical portion of the curve will now begin below pH 7 and end point lies between 4 and 6. This is again due to the hydrolysis of the salt which gives H^+ ions in aqueous solution.

(iv) *Titration of a weak acid against a weak base:* The nature of the curve depends on the ionisation constants of the weak acid (pK_a) and the weak base (pK_b) involved in the titration. If both the acid and base have very large pK_a and pK_b, the rise in pH at the end point is very small and may not be even observed.

Choice of an Indicator

The indicator selected for acid-base titrations should have the following characteristics:

(1) *It should change its colour at the pH, corresponding to the end point of the reaction, indicating that the reaction is complete.*

(2) *The colour change should be sharp.*

(3) *The change should be between the contrasting colours which are easily distinguished.*

Thus in order to choose an indicator, it is necessary to know the pH of a solution at the end point. For example, when a strong acid is titrated against a strong base or *vice-versa* the resulting solution is very nearly neutral at the end point.

Therefore, the ideal indicator will have a sharp colour change at about pH = 7. Thus any indicator which undergoes colour change within the range 4-10 can be used.

In the titration fo a weak acid with a strong base, the pH at the end point will be above 7 and any indicator changing colour between pH of 8 and 10 will be satisfactory. In the titration of a strong acid and weak base, the pH at the end point will be below 7 and any indicator, changing colour between pH 3 and 6 will be satisfactory.

In the titration of a weak acid with weak base, no sharp change in pH is obtained, so no indicator will give a satisfactory colour change.

This means the choice of indicator is limited as follows:

Table 2 : Choice of indicators

Titration	*Marked pH range*	*Indicator*
Strong acid and strong base	4—10	Any indicator
Weak acid and strong base	7.5—10.5	Phenolphthalein (8.3—10)
Strong acid and weak base	3.5—6.5	Methyl red (4.4—6.3) or Methyl orange (3.1—4.5)
Weak acid and weak base	No marked change	End point cannot be detected accurately by any indicator.

Universal Indicators

These are mostly employed for determining the approximate pH of solutions. It is a mixture of methyl red, methyl orange, bromothymol blue and phenolphthalein. It covers a pH range of 3–11 and gives colour changes at different pH values as shown below:

Colour	*Red*	*Orange*	*Yellow*	*Yellowish green*	*Blue*	*Violet*
pH	3	4-5	6	7	9	10

Buffer Solutions or Buffers

We know that the pH of neutral water is 7. If we add 1 cc of 1N HCl to 1 litre of water, the $[H^+]$ concentration in the solution becomes 10^{-3} gm ions and the pH of the solution becomes 3. Similarly, the addition of 1 cc of 1N NaOH in a litre of pure water changes the pH of the solution to 11. Thus, we see that the pH of water changes by large amounts on the addition of small quantity of a strong acid or a strong base. But for certain biological and analytical purposes, we require solution whose pH should not change on keeping or on addition of small amounts of a strong acid or a strong base. *Such solutions which oppose the change in their pH on the addition of small amount of an acid or a base are called buffer solutions.* Some characteristics of buffer solution are:

(i) Their pH should not change on keeping.

(ii) Their pH should remain same on dilution.

(iii) Their pH should remain same on the addition of small amounts of strong acid and base.

BUFFER CAPACITY

Often it is necessary to know the effectiveness of a buffer on a quantitative basis. To do so, we employ the term *buffer capacity* (β), first introduced by van Slyke in 1922. Buffer capacity is defined as the amount of acid or base that must be added to the buffer to produce a unit change of pH. Hence

$$\beta = \frac{d[B]}{d\,pH} \qquad ...(1)$$

where d[B] is the increase (in mol litre^{-1}) of strong base B. If a strong is added, Eq. (1) becomes

$$\beta = \frac{-d[B]}{-d\,pH}$$

The buffer capacity always has a positive value, however, since addition of base increases the pH and addition of acid decreases the pH. Thus, d[B] and dpH always have the same signs. The value of β depends not only on the nature of the buffer, but also on the pH, which is determined by the relative concentrations of the acid and its conjugate base. Plots of buffer capacity versus pH for the CH_3COOH—CH_3COONa system. We see that the buffer functions best around its pKa value of 4.76.

Action of Buffers

A buffer solution invariably consists of a weak acid and its salt solution (acidic buffer) or a weak base and its salt solution (basic buffer). Some examples of buffer solutions are:

Constituents	*pH Range*
Glycine + Glycine hydrochloride	1.0 — 3.7
Phthalic acid + Potassium hydrogen phthalate	2.2 — 3.8
Acetic acid + Sodium acetate	3.7 — 5.6
Disodium citrate + Trisodium citrate	5.0 — 6.3
Monosodium phosphate + Disodium phosphate	5.8 — 8.0
Boric acid + Borax	6.8 — 9.2

Borax + Sodium hydroxide	9.2 — 11.2
Disodium phosphate + Trisodium phosphate	11.0 — 12.0

Action of Acidic Buffer

Let us illustrate its action by considering a mixture sodium acetate and acetic acid in water.

$$CH_3COOH \rightleftharpoons CH_3COO^- + H^+ \text{ (feebly ionised)}$$
$$CH_3COONa \rightleftharpoons CH_3COO– + Na^+ \text{ (highly ionised)}$$

As CH_3COONa is almost fully ionised, the acetate ions so produced suppress the ionisation of the acetic acid, so that the mixture contains unionised acetic acid molecules and a large number of acetate ions. When a few drops of an acid, say HCl, are added the H^+ ions from the acid added react with the CH_3COO^- ions to form unionised or feebly ionised acetic acid:

$$CH_3COO^- + H^+ \rightarrow CH_3COOH$$

On the other hand, when a few drops of an alkali, say, NaOH are added, the OH^- ions react with CH_3COOH to form unionised H_2O.

$$CH_3COOH + OH^- \rightarrow CH_3COO^- + H_2O$$

Thus, the addition of small amounts of the acid or alkali to a buffer solution does not alter the H^+ ions concentration of the buffer. In other words, the pH of the buffer remains unchanged.

Action of a Basic-buffer

Let us illustrate the action of a basic buffer by considering a mixture of NH_4OH and NH_4Cl in water. They exhibit the following equilibria

$$NH_4OH \rightleftharpoons NH_4^+ + OH^- \text{ (weakly ionised)}$$

$$NH_4Cl \rightleftharpoons NH_4^+ + Cl^- \text{ (completely ionised)}$$

The presence of the excess of NH_4^+ ions from completely ionised NH_4Cl suppresses the ionisation of NH_4OH, so that the mixture solution contains unionised NH_4OH molecules and a large number of NH_4^+ ions. When a few drops of alkali is added to this buffer solution, the OH^- ions produced from the alkali react with NH_4^+ present in solution to form unionised NH_4OH.

$$NH_4^+ + OH^- \rightarrow NH_4OH$$

On adding a few drops of the acid say HCl, the H^+ ions provided by the acid react with NH_4OH to give unionised H_2O.

$$NH_4OH + H^+ \rightarrow NH_4^+ + H_2O$$

Thus, the addition of small amounts of the alkali or acid to the above solution does not alter its H^+ ion concentration. In other words, the pH of buffer remains unchanged.

Mathematical Expression for the pH of an Acidic Buffer

Consider a solution having a weak acid HA and its salt BA, with a strong base. The ionisation of the weak acid is represented as:

$$HA \rightleftharpoons H^+ + A^-$$

or $$K_a = \frac{[H^+][A^-]}{[HA]} \quad \text{or} \quad [H^+] = K_a \frac{[HA]}{[A^-]} \quad ...(1)$$

where K_a is the dissociation constant of the weak acid., The salt BA is almost completely ionised. Thus, high concentration of the A^- ions provided by almost complete ionisation of the salt BA suppresses the ionisation of the weak acid, so that it may be assumed that the free A^- ions are entirely due to the salt. Thus

$$[A^-] = [\text{Salt}] \text{ and } [HA] = [\text{Acid}] \quad ...(2)$$

Substituting Eq. (2) in (1), we get

$$[H^+] = K_a \frac{[\text{Acid}]}{[\text{Salt}]}$$

Taking logarithm of both sides, we get

$$\log [H^+] = \text{loh } K_a + \log \frac{[\text{Acid}]}{[\text{Salt}]}$$

or $$-\log [H^+] = -\text{loh } K_a - \log \frac{[\text{Acid}]}{[\text{Salt}]}$$

or $$pH = pK_a - \log \frac{[\text{Acid}]}{[\text{Salt}]} \quad ...(3)$$

$$= PK_a + \log \frac{[\text{Salt}]}{[\text{Acid}]}$$

where $pK_a = -\log K_a$. Equation (3) is known as Henderson-Hasselbalch equation.

MATHEMATICAL EXPRESSION FOR THE PH OF A BASIC BUFFER

Consider a weak base BOH and its salt AB. The dissociation of the weak base may be written as

$$BOH \rightarrow B^+ + OH^-$$

or $$K_a = \frac{[B^+][OH^-]}{[BOH]} \text{ or } [OH^-] = \frac{K_b[BOH]}{[B^+]} \quad ...(1)$$

where K_b is the dissociation constant of the weak base.

The high concentration of B^+ ions produced by the complete ionisation of the salt, which thus suppresses the ionisation of the weak base so that it may be assumed that free B^+ ions in the solution are entirely due to the salt. Thus,

$$[B^+] = \text{Salt and } [BOH] = \text{Base}$$

Thus, equation (1) may be put as

$$[OH^-] = K_b \frac{[Base]}{[Salt]}$$

Taking the logarithm of the above equation, we obtain

$$\log [OH^-] = \log K_b + \log [Base] - \log [Salt]$$

or $$-\log [OH^-] = -\log K_b - \log [Base] + \log [Salt]$$

or $$pOH = pK_b + \log \frac{[Salt]}{[Base]} \quad ...(2)$$

But pH = 14 – pOH, Equation (2) becomes as

$$pH = 14 - \left(pK_a + \log \frac{[Salt]}{[Base]} \right) \quad ...(3)$$

Now $[H^+][OH^-] = k_w = 10^{-14}$ or $\log k_w = -\log 10 = -14$

or $$14 = -\log k_w \quad ...(4)$$

From Eqs. (3) and (4), we get

$$pH = -\log k_w - \left(pK_b + \log \frac{[Salt]}{[Base]} \right)$$

$$= -\log k_w - \left(-\log K_b + \log \frac{[Salt]}{[Base]} \right)$$

$$= -\log \frac{k_w}{k_b} + \log \frac{[\text{Base}]}{[\text{Salt}]} \quad ...(5)$$

Hydrolysis

Pure water is neutral and is feebly ionised into H^+ and OH^- ions as:

$$H_2O \rightarrow H^+ + OH^-$$

The concentration of H^+ and OH^- ions is same. Now when a salt is dissolved in water, it ionises into ions, *e.g.*, the salt AB will ionise as:

$$AB \rightarrow A^+ + B^-$$

The ions of the salt may react with the ions of water to give

$$A^+ + OH^- \rightarrow AOH$$

$$B^- + H^+ \rightarrow HB$$

Now, the solution will be acidic or alkaline depending upon the relative strengths of the acid HB and the base AOH. If the acid is strong and the base is weak, then the solution will be acidic in nature. If the base is strong and the acid is weak, the solution will be basic in nature. Thus, this phenomenon in which a salt interacts with water to produce either an acidic or alkaline solution is termed as hydrolysis. Actually, hydrolysis involves the interaction of the ions of a salt with the water to produce H^+ or OH^- ions in solution. Thus, it may be defined as:

"*The interaction between the ions of salt and the ions of water to given free H^- or OH^- ions in solution.*"

Example of Hydrolysis

Salt may be divided into four categories:

(a) Salts of Strong Acid and Strong Bases

These salt, when dissolved in water, produce a strong acid and a strong base, so that the relative proportions of H^+ and OH^- ions in solution remain unchanged and the solution remains neutral. IN simple words, it can be said that the salts of strong acids and bases do not undergo hydrolysis. Let us consider the dissolution of sodium chloride in water.

$$NaCl \rightarrow Na^+ + Cl^-$$

$$H_2O \rightarrow OH^- + H^+$$

$$Na^+ + OH^- \rightarrow NaOH \text{ (highly ionised)}$$

$$H^+ + Cl^- \rightarrow HCl \text{ (highly ionised)}$$

But the acid, HCl and the base, NaOH being equally strong are almost completely ionised. Thus, the concentration of H^+ and OH^- is not altered and the solution remains neutral. Examples of such salts are NaCl, KCl, Na_2SO_4, K_2SO_4, etc.

(b) Salts of Weak Acids and Strong Bases

These salts (like Na_2CO_3; CH_3COONa, KCN, etc.), when dissolved in water undergo hydrolysis in water to give an alkaline solution. For example,

$$CH_3COONa \rightarrow CH_3COO^- + Na^+$$

$$H_2O \rightarrow H^+ + OH^-$$

$$CH_3COO^- + H^+ \rightarrow CH_3COOH$$

$$Na^+ + OH^- \rightarrow NaOH$$

NaOH being a strong base is almost completely ionised and the concentration of OH^- ions in solutions is not altered. However, CH_3COOH being a weak acid is feebly ionised.

Thus the concentration of H^+ ions in solution falls. Thus, the solution has more of OH^- ions than the H^+ ions. Hence, it is alkaline in nature.

(c) Salts of Strong Acids and Weak Bases

Examples are NH_4Cl, $CuSO_4$, etc. When such salts are dissolved in water, they produce acidic solutions. Consider the dissolving of ammonium chloride.

$$NH_4Cl \rightarrow NH_4^+ + Cl^-$$

$$H_2O \rightarrow H^+ + OH^-$$

$$NH_4^+ + OH^- \rightarrow NH_4OH$$

$$H^+ + Cl^- \rightarrow HCl$$

HCl being a strong acid is almost completely ionised and the concentration of the H^+ ions in solution is not affected. NH_4OH, however, is a weak base and feebly ionises to give only few OH^- ions in solution.

Thus, the solution has a large concentration of the H^+ ions that the OH^- ions. Hence the solution is acidic in nature.

(d) Salts of Weak Acid and Weak Bases

Examples are CH_3COONH_4, $(NH_4)_2CO_3$, etc. Such salts when dissolved in water yield acidic or basic or even neutral solutions, depending upon the nature of the acid and the base produced. Consider the dissolving of ammonium acetate in water.

$$CH_3COONH_4 \rightarrow CH_3COO^- + NH_4^+$$

$$H_2O \rightarrow H^+ + OH^-$$

$$CH_3COO^- + H^+ \rightarrow CH_3COOH$$

$$NH_4^+ OH^- \rightarrow NH_4OH$$

Both CH_3COOH and NH_4OH are weak and are equally but feebly ionised. Hence the solution contains practically the same number of the H^+ and OH^- ions. Therefore, the solution is almost neutral. In case of $(NH_4)_2CO_3$, however, the solution is slightly basic because NH_4OH is relatively a stronger base than the H_2CO_3 acid.

MATHEMATICAL TREATMENT OF HYDROLYSIS

(a) Salt of Strong Acids and Weak Bases:

(i) *Hydrolysis constant.* The hydrolysis of a salt of a strong acid and a weak base may be represented as:

$$A^+B^+ + H_2O \rightarrow HOH + B^- + H^+$$

or

$$A^+ + H_2O \rightarrow AOH + H^+$$

Applying the law of mass action to the above equilibrium,

$$K = \frac{[AOH][H^+]}{[A^+][H_2O]}$$

As water is taken in large excess, it means that its active mass remains practically constant. Therefore,

$$K[H_2O] = \frac{[AOH][H^+]}{A^+}$$

$$K_h = \frac{[AOH][H^+]}{A^+}, \text{ where } K_h = K[H_2O] \quad ...(1)$$

The constant K_h is called the hydrolysis constant of the salt.

(ii) *Relation between K_h, K_b and K_w.* The dissociation of the weak base produced in the hydrolysis of a strong acid and a weak base may be represented as:

$$AOH \rightarrow A^+ + OH^-$$

Applying the law of mass action,

$$K_b = \frac{[A^+][OH^-]}{[AOH]} \quad ...(2)$$

where K_b is the dissociation constant of the weak base.

Multiplying equations (1) and (2), we obtain

$$[H^+][OH^-] = K_h \times K_b$$

But $[H^+][OH^-] = K_w$, the ionic product of water

$\because$ $$K_w = K_h \times K_b$$

or $$K_h = \frac{k_w}{K_b} \quad ...(3)$$

From equation (3), it is evident that the *weaker the base, the greater will be the hydrolysis constant of the salt.*

(iii) *Degree of hydrolysis:* Degree of hydrolysis of salt may be defined as:

"The fraction of the total salt hydrolysed when the equilibrium has been established is called the degree of hydrolysis."

It is denoted by the letter '*h*'. In simple words, the extent to which the salt has been hydrolysed is known as the degree of hydrolysis. If *c* is the initial concentration of the salt in moles per litre and *h* is the degree of hydrolysis on the attainment of equilibrium, then the concentrations of various species will be represented as:

$$\underset{c(1-h)}{A^+} + \underset{\text{Excess}}{H_2O} \rightarrow \underset{ch}{AOH} + \underset{ch}{H^+} \quad ...(4)$$

Applying the law of mass action,

$$K_h = \frac{(ch)(ch)}{c(1-h)} \quad ...(5)$$

$$= \frac{ch^2}{(1-h)} = \frac{ch^2}{(1-h)}$$

If the base is very weak, the value of h is very small as compared to unity. It means that 1 – h in the equation (5) may be replaced by 1, *i.e.*,

$$K_h = ch^2$$

or $$h = \sqrt{\left(\frac{K_h}{c}\right)} \qquad ...(6)$$

From equation (3), we know

$$K_h = \frac{k_w}{K_c} \qquad ...(7)$$

Replacing the above value of K_h in equation (6), we obtain

$$h = \sqrt{\left(\frac{k_w}{cK_b}\right)} \qquad ...(8)$$

From the relation (8), it follows that the degree of hydrolysis of a salt of a weak base and a strong acid is:

(i) *directly proportional to the ionic product of water*

(ii) *inversely proportional to the value of K_b, and*

(iii) *inversely proportional to the value of c.*

From the above it follows that degree of hydrolysis increases if the base is weaker *i.e.*, if K_b is smaller, and also increases if the concentration decreases or the dilution increases.

As k_w increases much more than K_b, it means that the degree of hydrolysis at a given concentration increases with the increase in temperature.

(iv) Relation between pH, pK_b and c. From equation (4), it follows:

$$[H^+] = ch$$

$$= c\sqrt{\left(\frac{k_w}{cK_b}\right)} \qquad \text{[From equation (8)]}$$

$$= \sqrt{\left(\frac{ck_w}{K_b}\right)} \qquad ...(9)$$

Taking logarithms and changing the signs throughout, Eq. (9) becomes as:

$$-\log [H^+] = -\frac{1}{2}\log k_w + \frac{1}{2}\log K_b - \frac{1}{2}\log c$$

$$pH = \frac{1}{2}pk_w - pK_b - \frac{1}{2}\log c \qquad ...(10)$$

From equation (10), it follows that the pH of the solution must be less than $\frac{1}{2}$ pK_w and is less than 7.0 and so solutions of salts of the type under consideration will be acid solutions.

(b) Salts of Weak Acids and Strong Bases

(i) *Hydrolysis constant*: The hydrolysis of a salt of a weak acid and a strong base may be represented as:

$$(A^+ + B^-) + H_2O \rightarrow HB + A^+ + OH^-$$

$$B^- + H_2O \rightarrow HB + OH^-$$

Applying the law of mass action,

$$K = \frac{[OH^-][HB]}{[B^-][H_2O]} \qquad ...(11)$$

As water is taken in large excess, its active mass remains practically constant. Therefore,

$$K[H_2O] = \frac{[OH^-][HB]}{[B^-]}$$

or

$$K_h = \frac{[OH^-][HB]}{[B]} \qquad ...(12)$$

where K_h is known as the hydrolysis constant of the acid.

(ii) *Relation between K_h, K_a and k_w*: The dissociation of weak acid, HB, produced during hydrolysis may be represented as:

$$HB \rightleftharpoons H^+ + B$$

Applying the law of mass action,

$$K_a = \frac{[H^+][B^-]}{[HB^-]} \qquad ...(13)$$

where K_a is known as the dissociation constant of the weak acid.,

Multiplying Eq. (12) by (13), we get

$$[H^+][OH^-] = K_a \times K_h$$

But $[H^+][OH^-] = k_w$

$\therefore$ $k_w = K_a \times K_b$

or

$$K_b = \frac{k_w}{K_a} \qquad ...(14)$$

From eq. (14), it follows that the hydrolysis constant, K_h, of the salt varies inversely as the dissociation constant, K_a, of the weak acid. Therefore, the weaker the acid, the greater is the hydrolysis constant of the salt.

(iii) *Degree of hydrolysis:* If c is the initial concentration fo the salt in moles per litre and h is the degree of hydrolysis on the attainment of equilibrium, the concentration of various species may be represented as:

$$\underset{c(1-h)}{B^-} + \underset{\text{excess}}{H_2O} \rightarrow \underset{ch}{HB} + \underset{ch}{OH^-} \qquad \text{...(14A)}$$

Applying the law of mass action,

$$K_h = \frac{(ch)(ch)}{c(1-h)} = \frac{ch^3}{1-h} \qquad \text{...(15)}$$

If the acid is very weak, h is negligible as compared to unity, *i.e.*, $1 - h \approx 1$. Thus, equation (15) becomes as:

$$K_h = ch^2$$

or

$$h = \sqrt{\left(\frac{K_h}{c}\right)}$$

From equation (14), we have

$$K_h = \frac{k_w}{K_a}$$

$$\therefore \qquad h = \sqrt{\left(\frac{k_w}{cK_a}\right)} \qquad \text{...(16)}$$

From equation (16), it follows that weaker the acid, greater is h, *i.e.*, the degree of hydrolysis. As value of k_w is also increased by a rise in temp., it means that the degree of hydrolysis 'h' will considerably increase with rise of temperature It is also seen that degree of hydrolysis 'h' also increases if the concentration decreases *i.e.*, dilution increases.

(iv) *Relation between pH, pK_a and c:* From Eq. (14A), the concentration of $[OH^-]$ ion is given by

$$[OH^-] = ch$$

But

$$[H^+][OH^-] = k_w$$

$$[H^+] = \frac{k_w}{ch} \qquad \text{...(17)}$$

From equation (16), we have

$$h = \sqrt{\left(\frac{k_w}{cK_a}\right)}$$

Substituting the value of h in (17), we have

$$[H^+] = \frac{k_w}{c}\sqrt{\left(\frac{cK_a}{k_w}\right)} = \sqrt{\left(\frac{k_w K_a}{c}\right)} \qquad ...(18)$$

Taking logarithms and changing the signs throughout, equation (18) becomes as

$$-\log[H^+] = -\frac{1}{2}\log k_w - \frac{1}{2}\log K_a + \frac{1}{2}\log c$$

or
$$pH = pk_w + \frac{1}{2}pK_e + \frac{1}{2}\log c \qquad ...(19)$$

From equation (19), it follows that pH or alkalinity of a solution of the salt of a weak acid and strong base increases with decreasing and strength, *i.e.*, increasing pK_a and increasing concentration.

(c) Salt of Weak Acids and Weak Bases

(i) *Hydrolysis constant*: The hydrolysis of the salt ˛ f a weak acid and a weak base may be represented as:

$$AB + H_2O \rightleftharpoons AOH + HB$$

Ionically, the above equilibria can be represented as:

$$A^+ + B^- + H_2O \rightleftharpoons \underset{\text{weak base}}{AOH} + \underset{\text{weak acid}}{HB} \qquad ...(19A)$$

Applying the law of mass action

$$K = \frac{[AOH][HB]}{[A^+][B^-][H_2O]}$$

$$K[H_2O] = \frac{[AOH][HB]}{[A^+][B^-]}$$

$$K_h = \frac{[AOH][HB]}{[A^+][B^-]} \qquad ...(20)$$

where K_h is the hydrolysis constant of salt.

(ii) Relation between K_h, k_w, K_a, and K_b: The dissociation of the weak acid and weak base may be represented as:

$$HB \rightarrow H^+ + B^- \qquad ...(21)$$

and $$AOH \rightarrow A^+ + OH^- \qquad ...(22)$$

Applying the law of mass action to the above equilibria (21) and (22),

$$K_a = \frac{[H^+][B^-]}{[HB]} \qquad ...(23)$$

$$K_b = \frac{[A^+][OH^-]}{[AOH]} \qquad ...(24)$$

where K_a and K_b are the dissociation constants of the weak acid and base respectively. Multiplying Eqs. (20), (23) and (24),

$$[H^+][OH^-] = K_a \times K_b \times K_h$$

But $$[H^+][OH^-] = k_w$$

$$\therefore \quad K_w = K_a \times K_b \times K_h$$

or $$K_h = \frac{k_w}{K_a \times K_b} \qquad ...(25)$$

(iii) *Degree of hydrolysis:* If c is the initial concentration of the salt in moles/litre and 'h' is the degree of hydrolysis, then the concentration of various species of equilibria (19A) may be represented as:

$$\underset{c(1-h)}{A^+} + \underset{c(1-h)}{B^-} + \underset{\text{excess}}{H_2O} \rightarrow \underset{ch}{AOH} + \underset{ch}{HB} \qquad ...(26)$$

Applying the law of mass action,

$$K_h = \frac{[ch][ch]}{c(1-h)c(1-u)} = \frac{h^2}{(1-h)^2} \qquad ...(27)$$

If h being very small, h may be neglected and 1 – h may be taken as unity. Thus, equation (27) becomes as:

$$K_h = h^2 \qquad ...(28)$$

From Eq. (28) it follows that degree of hydrolysis of a weak acid and a weak base is independent of the dilution or concentration.

(iv) *Relation between pH, pK_w, pK_a and pK_b;* From equation (23), it follows that

$$[H^+] = K_a \frac{[HB]}{[B^-]} \qquad ...(29)$$

From equation (26), we get

$$B^- = c(1 - h) \text{ and } HB = ch \quad ...(30)$$

Substituting equation (30) in (29), we have

$$[H^+] = K_a \frac{ch}{c(1-h)} = K_a \frac{h}{1-h}$$

$$= K_a(K_h)^{1/2} \quad \text{[from equation (27)]}$$

$$= K_a\left(\frac{k_a}{K_a K_b}\right)^{1/2} \quad \text{[from equation (25)]}$$

$$= \left(\frac{k_w K_a}{K_b}\right)^{1/2}$$

$$= K_4(K_h)^{1/2} \quad \text{[from equation (27)]}$$

$$K_a = \left(\frac{K_w}{K_b}\right)^{1/2} \quad \text{[from equation (25)]}$$

$$= \left(\frac{k_w K_a}{K_a K_b}\right) \quad ...(31)$$

Taking logarithm of both sides obtain, we get

$$-\log [H^+] = -\frac{1}{2}\log k_w - \frac{1}{2}\log K_a + \frac{1}{2}\log K_b$$

$$pH = \frac{1}{2}pk_a + \frac{1}{2}pK_a - \frac{1}{2}pK_b \quad ...(32)$$

Now three different cases may arise:

(i) When $K_a = K_b$, it follows from equation (32), $pH = \frac{1}{2} pk_w = 7$, *i.e.*, the solution will be neutral.

(ii) When $K_a > K_b$; $pH < 7$, *i.e.*, the solution will be acidic.

(iii) When $K_a < K_b$; $pH > 7$, *i.e.*, the solution will be alkaline.

DETERMINATION OF DEGREE OF HYDROLYSIS

(a) From Colligative Properties

The colligative property of a solution number of the solute particles. Consider, 1 gm mole of the salt of weak acid and strong base like CH_3COONa which gets hydrolysed in the solution as:

$$\underset{1-x}{CH_3COO^-} + \underset{1-x}{Na^+} + H_2O \rightarrow \underset{x}{CH_2COOH} + \underset{x}{Na^+} + \underset{x}{OH^-}$$

where x is the degree of hydrolysis. Then,

total number of particles before hydrolysis = 1 + 1 = 2

and the total number of particles after hydrolysis

$$= 1 - x + 1 - x + x + x + x$$

$$= 2 + x$$

$$\because \frac{\text{Observed freezing point depression}}{\text{Calculated freezing point depression}} = \frac{2+x}{2}$$

From the above relation, x can be calculated.

(b) From Conductance Measurements

The method is based upon the simple assumption that the equivalent conductance of an aqueous solution of a salt of a weak acid and strong base (or a weak base and strong acid) is due to the sum of equivalent conductances of strong base (or a weak base and strong acid) is due to the sum of equivalent conductances of

(a) the unhydrolysed salt and

(b) the free OH^- (or the H^+) ions produced by hydrolysis.

The weak acid or weak base formed during hydrolysis do not contribute anything towards the equivalent conductance of the salt solution at infinite dilution.

Procedure: The procedure involves the following steps:

(i) An aqueous solution fo salt of known dilution is prepared. Now determine the equivalent conductance at a certain dilution, say, V, *i.e.* λ_V.

(ii) Now mix the aqueous solution of the salt with excess of the weak acid or weak base as the case may be or the same which is produced during hydrolysis. This addition of weak acid or weak base checks the hydrolysis of the salt. Now determine the equivalent conductance of the mixture at the same dilution as before. Let it be λ'_V. Actually, this is the equivalent conductance of the unhydrolysed salt.

(iii) The equivalent conductance of the strong base or acid at infinite dilution is also determined by Kohlrausch's law. Let it be λ_∞.

Calculation: From the above values of the various equivalent conductances, the degree of hydrolysis may be calculated as follows:

Consider a salt AB of a strong acid and a weak base. Its hydrolysis may be represented as:

$$A^+B^- + H_2O \rightleftharpoons AOH + H^+ + B^-$$

or

$$\underset{1-h}{A^+} + H_2O \rightleftharpoons \underset{h}{AOH} + \underset{h}{A^+}$$

If h is the degree of hydrolysis, then

Concentration f unhydrolysed salt = 1 – h

Concentration of free H^+ ions = h

Equivalent conductance due to unhydrolysed salt = $\lambda'_V - (1 - h)$ and equivalent conductance due to H^+ ions = $\lambda_\infty \times h$

∴ Equivalent conductance of hydrolysed salt,

$$\lambda_V = \lambda_\infty \times h + \lambda'_V(1 - h)$$

or

$$h = \frac{\lambda_V - \lambda'_V}{\lambda_\infty - \lambda'_V}$$

This is known as Bredig's method.

(c) pH Determination Method

In case of the hydrolysis of a salt of a strong acid and a weak base,

$$A^+ + H_2O \rightleftharpoons AOH + H^+$$

and the value of $[H^+]$ is given as

$$[H^+] = ch \qquad ...(1)$$

where 'C' is the concentration of the salt in gm moles/litre and h is the degree of hydrolysis.

Taking logarithm of both sides of equation (1), we obtain

$$\log [H^+] = \log ch$$

or

$$-\log [H^+] = -\log ch$$

or

$$pH = -\log ch \qquad ...(2)$$

In actual practice the aqueous solution of the salt is taken and its pH is determined with the help of pH meter. Thus, one can calculate the degree of hydrolysis.

(d) From the Dissociation Constants of the Weak Acids or Weak Bases

As we know in the case of

(i) a salt of a weak acid and a strong base

$$h = \sqrt{\left(\frac{k_w}{K_a \times C}\right)} \quad ...(1)$$

(ii) a salt of a weak base and a strong acid

$$h = \sqrt{\left(\frac{k_w}{K_b \times C}\right)} \quad ...(2)$$

(iii) a salt of a weak acid and a weak base

$$h = \sqrt{\left(\frac{k_w}{K_a K_b}\right)} \quad ...(3)$$

Knowing dissociation constant of acid or base (K_a or K_b) or both as the case may be, it is a simple matter to calculate the degree of hydrolysis of a given salt at a given concentration.

(e) Distribution Method

This method can also be used provided the free acid or base formed during hydrolysis is soluble in a immiscible solvent like benzene. Let us illustrate this principle by considering the hydrolysis of aniline hydrochloride.

When aniline hydrochloride is hydrolysed, the following equilibria are established in solution.

$$C_6H_5NH_3^+Cl^- + H_2O \rightarrow C_6H_5NH_3^+OH^- + HCl$$

$$\downarrow$$

$$C_6H_5NH_2 + H_2O$$

Applying the law of mass action, the hydrolysis constant K_b is given as:

$$K_h = \frac{[\text{Concentration of free base}]\,[\text{Concentration of free acid}]}{[\text{Concentration of unhydrolysed salt}]}$$

If c is the initial concentration of the salt in moles per litre, then the degree of hydrolysis is given by:

$$h = \left(\frac{K_h}{c}\right)^{1/2} \qquad ...(1)$$

The concentration of aniline formed by hydrolysis of aniline hydrochloride can be determined as follows:

(i) A known amount of aniline is distributed between benzene and water. The concentration of aniline in the benzene layer is measured by passing dry HCl gas through it and weighing the amount of aniline 'hydrochloride precipitated. Difference from the total weight of aniline taken gives the concentration of aniline in water layer. When distribution law is applied, distribution coefficient 'K' is given by

$$K = \frac{\text{Aniline in benzene}}{\text{Aniline in water}}$$

(ii) Now shake a known weight of aniline hydrochloride with a known weight of benzene and water. The free aniline formed distributes itself between benzene and water. Amount of the aniline in the benzene layer is determined by precipitating as aniline hydrochloride by passing dry HCl gas. Then, calculate the amount of aniline in the water layer by using the distribution coefficient determined in the first step.

The sum of the weights of aniline in the benzene and water layers gives the total weights of aniline formed by hydrolysis. The concentration of free HCl (in moles per litre) will be the same as of aniline in moles/litre.

The concentration of unhydrolysed salt (c)

= [Initial concentration of the salt taken]

– [the loss of concentration due to hydrolysis]

= [Initial concentration of the salt]

– [Concentration of aniline formed by hyrolysis]

Substituting the various values in (1), the hydrolysis constant or the degree of hydrolysis of aniline hydrochloride can be determined.

TRANSPORT NUMBER

The fraction of the total current carried by any one of the ionic species is known as transport number, transference number or Hittorf number of the species and may be denoted by sets of symbols like t_+ and t_-, t_a and t_c or n_c are n_a.

From the above definition, it follows that the transport number of an anion is

$$t_- = \frac{\text{Current carried by an anion}}{\text{Total current passes through the solution}} \quad ...(1)$$

and the transport number of cation is

$$t_+ = \frac{\text{Current carried by the cation}}{\text{Total current passed through the solution}} \quad ...(2)$$

But we know that the quantity of electricity carried by each ion is directly proportional to the speed of the concerned ions. So, amount of electricity carried by anion $\propto$ speed of anion (u_a) amount of electricity carried by cation $\propto$ speed of cation (u_c).

$\because$ Total amount of electricity carried $\propto$ speed of anion + speed of cation

$$\propto u_a + u_c$$

Thus, the transport number of anion [From eq. (1), t_- is]

$$t_- = \frac{u_a}{u_a + u_c} \quad ...(3)$$

and the transport number of cation [From eq. (2)], t_+, is

$$t_+ = \frac{u_c}{u_a + u_c} \quad ...(4)$$

It follows from equation (3) and (4) that the transport number of an ion may be defined as the ratio of speed of the ion to the sum of the speeds of both cation and anion.

By adding Eqs. (3) and (4), we get

$$t_- + t_+ = \frac{u_a}{u_a + u_c} + \frac{u_c}{u_a + u_c} = \frac{u_a + u_c}{u_a + u_c} = 1 \quad ...(5)$$

$$t_- + t_+ = 1$$

From eq. (5), it follows that

(i) The sum of the transport numbers of all the species taking part in the transport of electricity is equal to unity, and

(ii) If the transport number of one of the ionic species is known, then that of the other can be easily calculated by using eqn. (5).

Experimental Determination of Transport Number

There are several methods for determining the transport numbers of ions, viz.

(i) Hittorf's Method

Principle : This method is based on the principle that "*During electrolysis, the fall of concentration around the electrode is proportional to the speed of ions moving away from that electrode.*"

Thus, the fall of concentration around the cathode $\propto$ speed of the anion

$$(u_a), \qquad \text{...(6)}$$

and the fall of concentration around the anode $\propto$ speed of the cation

$$(u_c) \qquad \text{...(7)}$$

Dividing eq. (6) by (7), we get

$$\frac{\text{Fall of concentration around the cathode}}{\text{Fall of concentration around the anode}} = \frac{u_a}{u_c}$$

Adding 1 to both sides, we get

$$1 + \frac{\text{Fall of concentration around the cathode}}{\text{Fall of concentration around the anode}} = 1 + \frac{u_a}{u_c}$$

or

$$\frac{\text{Fall of concentration around the anode + Fall of concentration around the cathode}}{\text{Fall of concentration around the anode}} = \frac{u_c + u_a}{u_c}$$

or

$$\frac{\text{Total fall of concentration}}{\text{Fall of concentration around the anode}} = \frac{u_c + u_a}{u_c}$$

Taking reciprocals of the abode equation,

$$\frac{\text{Fall of concentration around the anode}}{\text{Total fall of concentration}} = \frac{u_c}{u_c + u_a} \qquad \text{...(8)}$$

From equation (4), we know $t_+ = \dfrac{u_c}{u_a + u_c}$

$$\therefore \frac{\text{Fall of concentration around the anode}}{\text{Total fall of concentration}} = t_+ \qquad \text{...(9)}$$

Similarly,

$$\frac{\text{Fall of concentration around the cathode}}{\text{Total fall of concentration}} = t_- \qquad \text{...(10)}$$

Discussion of the Principle

(i) Equations (9) and (10) are valid only if the electrodes are not attacked by the ions in solution.

(ii) If the anode is attacked by the anions, there will be an increase in the concentration around the anode instead of decrease. However, this increase in concentration at anode is equal to decrease in concentration that would have taken place if the anode were inert. Example is in the case of electrolysis of $AgNO_3$ using Ag anode.

Apparatus: Suppose one wants to determine the transport number of Ag^+ and NO_3^- ions in a solution of silver nitrate.

(i) It consists of two vertical glass tubes connected by a V-tube in the middle. The end tubes containing the anode and cathode constitute the anodic and cathodic compartments respectively and V-tube constitutes the central compartment.

(ii) The tubes are provides with stop cocks at the bottom.

(iii) The electrodes are made of a suitable metal (attackable or non-attackable) and sealed in glass tubes which pass through rubber stoppers. Some mercury is also placed in the glass tubes to ensure proper contact.

(iv) The apparatus is filled up with a standard solution of silver nitrate.

(v) In addition to this apparatus, the circuit also contains a silver voltameter or an ammeter for determining the total electricity passed. The voltameter contains standard $AgNO_3$ solution and platinum electrodes.

Working

(i) A dilute solution fo known concentration (N/10) of the salt containing the ions, of which the transport numbers are to be determined, is filled in the apparatus.

(ii) The apparatus is then connected to a battery, variable resistance and voltameter in series. A low current of the order of 1 – 20 milliamperes is passed for 2 to 3 hours.

If high current density is used, diffusion sets in and the solution sin the various compartments are mixed and thus leading to wrong results.

(iii) After the electrolysis the stop cocks connecting the U-tube with the side tubes are closed and the solutions are withdrawn from the various compartments separately. The concentration is determined by titration against KCNS solution. These should be no change in the concentration of the solution from the V-tube, *i.e.*, central compartment. Knowing the original concentration of the solution taken, the change in concentration is determined separately.

(iv) The weight of Ag(or Cu) deposited in the voltameter is also determined separately. This represents the total quantity of electricity passed during electrolysis.

Calculations: Calculation involves the following steps:

Case I : *When the electrodes are unattackable.* This condition can be obtained by using platinum electrodes. Suppose that,

(i) The weight of $AgNO_3$ in the a g of anodic solution before electrolysis = X gm.

(ii) The weight of $AgNO_3$ in the b g of anodic solution after electrolysis = Y gm.

(iii) The weight of $AgNO_3$ in the a g of anodic solution after electrolysis

$$= \frac{Y}{b} \times a = Z$$

∴ Full in concentration around anode = (X – Z) gm.

Let the weight of Ag deposited in the voltameter = 2 gm

$$\frac{\text{Total weight of AgNO}_3\text{ electrolysed}}{\text{Weight of Ag deposited in voltameter}}$$

$$= \frac{\text{Eq. wt. of AgNO}_3}{\text{Eq. wt. of Ag}}$$

∴ $$\frac{\text{Total weight of AgNO}_3\text{ electrode}}{w} = \frac{170}{108}$$

or Total weight of $AgNO_3$ electrolysed $= \dfrac{w \times 170}{108} = W$ (say)

Now the transport No. of Ag^+ ion is given by

$$t_{Ag^+} = \frac{\text{Loss in wt. of AgNO}_3\text{ around anode}}{\text{Total weight of AgNO}_3\text{ electrolysed}}$$

$$= \frac{X - Z}{W} \qquad ...(11)$$

and the transport No. of NO_3^- ions is given by

$$t_{NO_3^-} = 1 - t_{Ag^+} = 1 - \frac{X - Z}{W} \qquad ...(12)$$

Case II : *When the electrodes are attackable.* This condition can be obtained by using Ag anode in $AgNO_3$ solution.

In this case $Z \geq X$, *i.e.*, the concentration of $AgNO_3$ in the anodic compartment after electrolysis is more than that before electrolysis. This is due to the fact that NO_3^- ions after losing charge at the anode attack the silver anode to form $AgNO_3$. Thus, the concentration of $AgNO_3$ in the anodic compartment increases.

The fall in concentration of $AgNO_3$ in the anodic compartment due to the migration of the Ag^+ ions may be calculated as follows:

The amount of $AgNO_3$ formed by the discharge of NO_3^- ions at the anode = No. of the NO_3^- ions discharged.

= total $AgNO_3$ electrolysed W (say)

∴ The actual increase in conc. of anodic compartment = W.

But observed increase in conc. of anodic compartment = Z – X.

∴ Fall in concentration due to the migration of Ag^+ ions

= Actual increase – observed increase

= W – (Z – X)

Hence
$$t_{Ag^+} = \frac{W - (Z - X)}{W} \qquad ...(13)$$

and
$$t_{NO_3^-} = 1 - t_{Ag^+} = 1 - \frac{W - (Z - X)}{W} = \frac{Z - X}{W} \qquad ...(14)$$

(2) Moving Boundary Method

Principle: It is based on measuring the rate of migration of one or both of the ionic species of the electrolyte, away from the similarly charged electrodes. This method was first of all suggested by Lodge and Whetham. It was modified by Masson (1899), Steel (1904) and Denison (1906). Finally it was the work of Macinnes (1923) whose work made the procedure more easy and simple.

(i) Suppose one is interested in measuring the transport number of any cation, say Na^+ in NaCl. To do it, one has to select another suitable electrolyte called the *indicator-electrolyte* which has a common ion with the electrolyte under investigation. It is the usual practice to select such an electrolyte as indicator electrolyte whose cation must be slow as compared to the cation of the electrolyte of which one is interested in determining the transport number. For example, LiCl is used as an indicator electrolyte while determining the transport number of Na^+ because Li^+ ions move at a slower rate than Na^+. If one takes indicator ion moving faster than the experimental ion, the boundary line gets blurred. In most of the methods, LiCl is not used as an indicator but instead of this cadmium is used as an anode which on electrolysis yields cadmium chloride which in turn itself acts as an indicator electrolyte.

(ii) The electrolytic cell consists of the shape as shown, the middle part of the left hand portion is vertical tube of uniform area of cross section.

(iii) The experimental solution NaCl is floated on top of a denser solution of $CdCl_2$.
At the bottom of the tube is fixed an anode of cadmium metal. Over the solution of NaCl a less dense solution of sodium acetate is floated.

(iv) The cathode used is of platinum. This is placed in a side tube so that the evolution of hydrogen gas at the cathode does not disturb the main liquid column.

Working

(i) A constant current is passed for 5 to 6 hours. The Na^+ ions move upwards towards the cathode closely followed by the Cd^{2+} ions. The sharp boundary moves gradually upwards. The movement of the boundary between NaCl and $CdCl_2$ can be easily followed on account of a difference in the refractive indices of the two solutions.

(ii) The rate of migration of Cl^- ions can also be observed as the acetate anions are slow moving than the chloride ions.

(iii) At the end of experiment, the distance through which the boundary has moved is noted. The time for which the current has passed is noted.

Calculations: Suppose a current of I amperes is passed for t seconds, the quality of electricity carried by Na^+ ions is given by

$$= t_+ \text{ It coulombs}$$

where t_+ is the transport number of the Na^+ cation

Suppose the concentration of NaCl solution = c gm eq. eq/c.c.

Suppose the area of cross section of the moving boundary = a cm^2.

Therefore, the amount of Na^+ ion that has migrated upwards

$$= \frac{t_+ It}{F} \text{ where F is Faraday.} \qquad ...(15)$$

Distance by which NaCl—$CdCl_2$ boundary moving during t seconds = x cm.

Volume of the solution cleared by the migration of Na^+ ions = ax

Number of gm equivalent of Na^+ ions transported = axc ...(16)

From Eqs. (15) and (16), we have $axc = \frac{t_+ It}{F}$

or
$$t_+ = \frac{F\, axc}{It} \qquad ...(17)$$

Similarly,
$$t_- = \frac{F \cdot ayc}{It} \qquad ...(18)$$

where y is the distance by which NaCl—CH_3COONa boundary moves during t send and t_- is the transport number of chloride anion.

Also
$$\frac{t_+}{t_-} = \frac{x}{y}$$

But
$$t_- = 1 - t_+$$

$\therefore$
$$\frac{t_+}{1 - t_+} = \frac{x}{y} \text{ or } t_+ = \frac{x}{x + y} \qquad ...(19)$$

Similarly
$$t_- = \frac{x}{x + y} \qquad ...(20)$$

Thus, t_+ and t_- can be evaluated.

(3) E.M.F. Method

Please see Application of E.M.F. measurements.

(4) From Ionic Mobility

The ionic mobility or conductance of an ion is directly proportional to the speed of the ion. Thus,

$$\lambda_+ \propto u_+$$

or
$$\lambda_+ = Ku_+ \qquad ...(21)$$

Similarly,
$$\lambda_- \propto u_-$$

or
$$\lambda_- = Ku_- \qquad ...(22)$$

In Eqs. (21) and (22), u_+ and u_- are the speeds while λ_+ and λ_- are the ionic conductivities of the cation and anion respectively.

Now the transport number of cation is given by

$$t_+ = \frac{u_+}{u_+ + u_-}$$

$$= \frac{\frac{1}{K}\lambda_+}{\frac{1}{K}(\lambda_+ + \lambda_-)} \quad \text{[From Eqs. (21) and (22)]}$$

$$= \frac{\lambda}{\lambda_+ + \lambda_-} \quad \text{...(23)}$$

But $\lambda_+ + \lambda = \lambda_\infty$ (Kohlrausch's Law)

$$\therefore \quad t_+ = \frac{\lambda_+}{\lambda_\infty} \quad \text{...(24)}$$

Similarly

$$t_- = \frac{\lambda_-}{\lambda_\infty} \quad \text{...(25)}$$

From Eqs. (24) and (25), it follows that the transport number of cation and anion can be determined if one knows the values of λ_+, λ and λ_∞.

In this case, oxidation takes place at the zinc electrode and reduction at the hydrogen electrode, as shown below:

Zinc electrode $Zn\ (s) \rightleftharpoons Zn^{2+} + 2e^-$

Hydrogen electrode $2H^+ + 2e^- \rightleftharpoons H_2(g)$

$\therefore$ *The net reaction is* $Zn(s) + 2H^+ \rightleftharpoons Zn^{2+} + H_2(g)$ (1 atm)

NERNST'S EQUATION FOR SINGLE ELECTRODE POTENTIAL

From the Nernst théory it follows that the potential difference of the double layer formed at one electrode is termed as single electrode potential.

Let us consider a metal of valency 'n' dipping in a solution

$$M \rightleftharpoons M^{+n} + ne$$

Let P_1 be the osmotic pressure of its ions, P_3, be the solution pressure of the metal and E be the electrode potential, *i.e.* the difference of potential between the metal and solution. Now pass the current through the electrode reversible until 1 gm. ion of the metal gets dissolved. It means that the electricity required for the dissolution of 1 gm. ion of the metal will be nF* coulombs. Hence, the electrical work done = NF*E volt-coulombs

i.e. $w_1 = nF^*E$ volt-coulombs

where F* is one faraday whose value is equal to 96500 coulombs and w_1 is the electrical work done .Go on diluting the solution till the osmotic pressure gets reduced from P_1 to $P_1 - dP_1$. At the same time the difference of potential between the metal and the solution is now changed from E to E – dE. At the potential difference E – dE, the electrical work done (w_2) to cause the dissolution of 1 gm. ion of the metal is given by

$$w_2 = (E - dE)\, nF^* \text{ volt-coulombs}$$

The difference in electrical energy of work = $w_1 - w_2$

or
$$w_1 - w_2 = nEF^* - (E - dE)nF^*$$
$$= nF^*\, dE \text{ volt coulombs.}$$

The different in electrical energy is equal to the osmotic work done in transferring 1 gm. ion of the metal from P1 to P_1—dP_1 which is equal to the product of volume and change in pressure *i.e.*,

$$nF^*dE \rightleftharpoons VdP_1 \qquad \text{...(1)}$$

where V is the volume of the solvent containing 1 gm. ion of the metal. For ideal solutions, osmotic pressures are directly proportional to the activities of ions, then $V = RT/P_1$. Thus, equation (2) reduces to

$$nF^*dE = \frac{RT\,dP_1}{P_1} \qquad \text{...(2)}$$

$$nF^* \int dE = RT \int \frac{dP_1}{P_1}$$

$$nF^*E = 2.303\ RT \log P_1 + K \qquad \text{...(3)}$$

where 'K' is the integration constant whose value will be evaluated as follows: when the solution pressure (P_2) is equal to osmotic pressure (P_1) of the ions, it means that no potential difference exists between the metal and the solution, *i.e.*, E = 0 and $P_1 = P_2$. Hence equation (3) becomes as

$$0 = 2.303\ RT \log P_2 + K$$

or
$$K = -2.303\ RT \log P_2 \qquad \text{...(4)}$$

Substituting the value of K from eq. (4) in (3), we get

$$nF^*E = 2.303\ (RT \log P_1 - RT \log P_2)$$

$$E = 2.303 \frac{RT}{nF^*} \log \frac{P_1}{P_2}$$

Equation (5) is termed as the Nernst equation for electrode potential. Let us now consider tow similar electrodes dipping in two different solutions. In this case, the potential difference between the electrodes will be given by

$$E_1 - E_2 = \frac{2.303\,RT}{nF^*}\left(\log\frac{(P_1)_A}{P_2} - \log\frac{(P_1)_B}{P_2}\right)$$

$$= \frac{2.303\,RT}{nF^*}\log\frac{(P_1)_A}{(P_2)_B} \qquad ...(6)$$

where $(P_1)_A$ and $(P_1)_B$ denote the osmotic pressure of the metal ions in two solutions A and B respectively. As we know that the osmotic pressure is proportional to concentration, it means, that equation (6) may be written as

$$E_1 - E_2 = \frac{2.303\,RT}{nF}\log\frac{(c_1)_A}{(c_2)_B}$$

Migration of Ions

Take a U-tube of uniform cross-section. Fill the lower limb of A, bend and the greater portion of the limb B with hot 4% solution of gelatin containing a few cc's of saturated KCl (to make it conducting) and a few drops of KOH plus a drop of phenolphathalein (to make it pink coloured). Allow it to cool, when it sets to pink coloured jelly. Then fill about 3/4th of the remaining portion of limb A with hot solution of gelatin + KCl + a drop of phenolphthalein. Allow it to cool when it sets to form a colourless jelly. Then fill the remaining upper portions of the limb A and B with KOH and HCl solutions respectively. Finally, dip in these solutions two platinised platinum electrodes in such a way that – ve electrode is in contact with KOH solutions, while + ve electrode is in contact with HCl solution. On passing electric current, the H^+ ions move towards cathode; while OH^- ions move towards anode.

The movement of these ions is indicated by decolorisation of pink colour in limb B and appearance of pink colour in limb A respectively. From the lengths of changed columns, due to the movement of H^+ and OH^- ions, it is found that speed of H^+ ions is nearly double that of OH^- ions. *Hence, it proves that different types of ions move with different speeds.*

Speeds of Ions and Amounts Liberated at the Electrodes

When an electrolyte is submitted to slow electrolysis using non attackable electrodes it is found that *two types of ions are discharged in equivalent amounts,* but *loss of electrolyte around the two electrodes is not the same.* This clearly shows that *oppositely charged ions migrate with different speeds* and the *ions which migrate with greater speed produce a greater loss in amount of electrolyte around the electrode it leaves.* The same may theoretically be illustrated as follows:

(i) Consider an electrolytic cell divided into three imaginary compartments as presented by dotted lines. Let us suppose that there are 4 pairs each of cations and anions in anodic and cathodic compartments; while 6 pairs in the middle compartment before electrolysis.

(ii) *Let only 2 anions move* from the cathodic to anodic compartment. The surplus unpaired ions in each compartment are two, which are discharged. As a result of this the *loss of number of molecules in the cathodic compartment is 2 molecules* (*i.e.*, 2 + ions and 2 – ions), while there is *no change in the anodic compartment.*

(iii) *Let cations and anions move with same speed:* Suppose in a particular time 2 anions and 2 cations move towards opposite electrodes. The discharge of cations and anions at the two electrodes is again the *same* (*i.e.*, 4 ions each) and the *loss of electrolyte in the two compartments is also same*, *i.e.*, 2 molecules each.

(iv) *Let cations move at double the speed of anions:* Suppose in a particular time one anion and two cations move to the opposite electrodes. Again, the total number of ions discharged in each compartment is *same*, *i.e.*, 3 ions each.

However, the *loss of electrolyte around the anode* (*i.e.*, 2 molecules) *is double the loss of electrolyte around the cathode* (*i.e.*, 2 molecules).

For the above it is clear that:

(a) *Loss of electrolyte around an electrode is proportional to the speed of the ion leaving it i.e.*,

$$\frac{\text{Loss around cathode}}{\text{Loss around anode}} = \frac{\text{Speed of cation}}{\text{Speed of anion}}$$

(b) *Whatever be the relative speeds of the two ions, the discharge of ions at the two electrodes is always equal.*

Standard Electrode Potential

When all the substances taking part in a reaction in a reversible cell are maintaining at unit activity, *i.e.*, in their standard states, the E.M.F. of any electrode in these conditions is known as standard electrode potential of the given cell (E°). If a reaction occurs by the passage of n Faradays in a cell in which the reactant and product have unit activities, the standard free energy change ΔG° is equal to $-$ nE F*. Hence

$$-\Delta G^\circ = nE^\circ F^* \quad ...(1)$$

where F* is the Faraday of electricity. Let us now consider a reversible process (chemical reaction between A and B to form the products L and M)

$$aA + bB + ... \rightleftharpoons lL + mM + ...$$

When we apply the law of mass action to the above equilibrium, we get

$$K = \frac{a_L^l a_M^m}{a_A^a a_B^b \cdots} \quad ...(2)$$

As the increase in free energy ΔG° at constant temperature is governed by the Vant Hoff's reaction isotherm, it, therefore, follows from this isotherm that

$$-\Delta G = RT \log_e K - RT \log_e \frac{a_L^l a_M^m}{a_A^a a_B^b \cdots} \quad ...(3)$$

If in the process the reactants and products are taken in their respective standard states, *i.e.*, at unity, then eq. (3) becomes as

$$-\Delta G^\circ = RT \log_e K \quad ...(4)$$

Eliminating ΔG° between equations (1) and (4), we get

$$nE^\circ F^\circ = RT \log_e K \quad ...(5)$$

If the concentrations of reactants and products are taken at any arbitrary activities, then E.M.F. of the cell (E) is given by

$$\Delta G = -nEF^* \quad ...(6)$$

Substituting equations (6) and (5) in (3), we obtain

$$nEF^* = nE^\circ F^* - RT \log_e \frac{a_L^l a_M^m}{a_A^a a_B^b \cdots}$$

$$E = E^{o} - \frac{RT}{nF*}\log_{e}\frac{a_L^l\, a_M^m}{a_A^a\, a_B^b \ldots} \qquad \ldots(7)$$

Expression (7) is the general equation for the E.M.F. of any reversible cell in which the reactants and products are taken at any arbitrary activities. If a_A, a_B, a_C, ... and a_L, a_M, a_N, ... are unity, it means that equation (7) becomes as

$$E^{o} = E$$

So normal or standard electrode potential may be defined as:

"*The potential of the electrode when the activities of the reactants and products are unity.*"

The earliest tabulation of standard electrode potential was mainly performed by Wilsmore and later revised by Abegg, Luther and Auerbach (1911). However, Lewis and Randall recalculated the standard electrode potential from activity data. Let us now consider a metal of valency z_+, reversible with respect to M^{z+} ions, the electrode reaction is

$$M \rightarrow M^{z+} + z_+e^-$$

The equation for the electrode potential is taken as:

$$E_+ = E^{o} - \frac{RT}{z_+F*}\log_e\frac{a_M^+}{a_M} \qquad \ldots(9)$$

where a_M^+ is the activity of the cations in the solution and a_M is the activity of the solid metal. As the solution state of metal is generally taken as unity, it means that equation (9) takes the form

$$E_+ = E^{o} - \frac{RT}{z_+F*}\log_e a_M^+ \qquad \ldots(10)$$

Again consider an electrode A which is reversible to A^- ions. The electrode reaction is

$$M^{z-} \rightarrow A + z^-e^-$$

Proceeding in the same manner as we have done earlier, we can prove that

$$E_- = E^{o} + \frac{RT}{z_-F*}\log_e a_A^- \qquad \ldots(11)$$

From expressions (10) and (11), it follows that the general expression for any electrode reversible with respect to single ion of valency $z\pm$ may be given as:

$$E_{\pm} = E^{\circ} \pm \frac{RT}{z \pm F^*} \ln a_l \qquad ...(12)$$

where a_l is the activity of a particular species. In equation (12), the upper signs are taken for a positive ion while the lower ones for a negative ion. Putting R = 8.313 Joules, F* = 96500 coulombs and by converting Neperian to Briggsian logarithms, the factor 2.3026 is utilised, it means that equation (12) becomes as

$$E_{\pm} = E^{\circ} \pm 1.9835 \times 10^{-4} \frac{T}{z\pm} \log a_l$$

$$= E^{\circ} \pm \frac{0.05915}{z\pm} \log a_l \text{ [when T = 298. 16°K]} ...(13)$$

Effect of concentration and valence on electrode potential. We have already proved that

$$E_l = E^{\circ} + \frac{0.0591}{n} \log a_l$$

From the above expression it is evident that a ten fold change in activity of an ion causes the electrode potential change by 0.0591/z± volts at 25°C. In a generalised manner, a change of activity by a factor of 10* changes the potential by $x \times 0.0591/z\pm$ volts at 25°. For monovalent ions, a ten fold change in concentration of solution will cause a change of 0.591 volts in monovalent and for bivalent ions, a change of 0.591/2 volts at 25°.

Sign of Electrode Potential

For years two different signal conventions for electrode potentials have been in common use.

Without real justification, they are usually referred to as

(i) *European convention.*

(ii) *American convention.*

American convention was designated by Lewis and Randall and has found popularity among the physical chemists. To illustrate the two conventions, let us consider the following reaction taking place at the zinc electrode:

$$Zn \rightarrow Zn^{2+} + 2e^-$$

According to European convention, one would assign a value of – 0.763 V for the standard potential of this electrode with respect to hydrogen electrode. On the other hand, the value according to the American convention, would be + 0.763 V. Thus, it would appear that the difference between the two conventions is the sign only. Let us now consider the reverse reaction f zinc electrode as:

$$An^{2+} + 2e^- \rightarrow Zn$$

According to the European convention, the standard electrode potential will be the same, *i.e.*, – 0.763 V but according to the American convention it is not + 0.763 V but is – 0.763. Thus, according to American convention it is evident that

(i) If the reaction is an oxidation, the electrode potential is positive as in $Zn \rightarrow Zn^{2+} + 2e^-$ and

(ii) If it is a reduction, the electrode potential is negative in $Zn^{2+} + 2e^- \rightarrow Zn$.

Thus, the electrode potential is a *bivariant* quantity according to the American convention, *i.e.*, both oxidation and reduction, whereas in the European convention it is an invariant quantity.

In an attempt to reconcile the two conventions, IUPAC have recommended changes in terminology that

(i) Electrode potential will be reserved for the European convention and,

(ii) E.M.F. of a half cell, in the American convention.

According to *latest convention* adopted by IUPAC (International Union of Pure and Applied Chemistry), *the electrode potential is given a positive sign if the electrode reaction involves* reduction (*i.e.* taking up of electrodes from the electrode) *when connected to the standard hydrogen electrode and a* negative sign *if the electrode reaction involves* oxidation (*i.e. liberation of electrons*) *when connected to the standard hydrogen electrode taken arbitrarily as zero.*

Thus, when copper electrode (copper rod dipping in a solution of a copper salt) is connected with a standard hydrogen electrode, *reduction* takes place at the copper electrode.

The electrode reactions are represented below:

Hydrogen electrode $\quad H_2\,(g) \rightleftharpoons 2H^+ + 2e^-$ (Oxidation)

Copper electrode $\quad Cu^{2+} + 2e^- \rightleftharpoons Cu(s)$ (Reduction)

Here, according to the above convention, the potential of copper electrode is taken as *positive*. Thus, $E(Cu^{2+}, Cu)$ is positive. However, if zinc electrode is connected with the standard hydrogen electrode, *oxidation* takes place at the zinc electrode. The electrode reaction are represented below:

Zinc electrode $Zn(s) \rightleftharpoons Zn^{2+} + 2e^-$ (Oxidation)

Hydrogen electrode $2H^+ + 2e^- \rightleftharpoons H_2(g)$ (Reduction)

Hence, the potential of the zinc electrode is taken as *negative*. Thus, $E(Zn^{2+}, Zn)$ is negative.

SOME SOLVED PROBLEMS

Problem 1:

The ionization constant of HCN = 4 × 10 [10]. Calculate the concentration of hydrogen ions in

(i) 0.2 M solution of HCN and

(ii) 0.2 M solution of HCN containing 1 mole/litre of KCN.

Solution:

(i) For a weak electrolyte like HCN, the degree of dissociation (α) is given by Ostwald's dilution law as:

$$a = \sqrt{K_a \times V}$$

Here $K_a = 4 \times 10^{-10}$; $V = 1/0.2 = 5$ litres

$$\therefore \quad \alpha = \sqrt{4} \times 10^{-10} \times 5 = 4.47 \times 10^{-5}$$

$$\therefore \text{ conc. of } [H^+] = \frac{\alpha}{V} = \frac{4.47 \times 10^{-5}}{5} = 8.94 \times 10^{-5} \text{ g-ion/litre}$$

(ii) When 1 mole/litre of KCN (a strong electrolyte,) is present, the concentration of CN^- ion $[CN^-] = 1.0$ g ion/litre (the conc. of CN^- from HCN being very small comparatively, and hence neglected); conc. of HCN = 0.2 g mole/litre.

$\therefore$ On substitution of these values in the expression:

$$K_a = \frac{[H^+][CN^-]}{[HCN]}$$

We get, $4 \times 10^{-10} = \dfrac{[H^+] \times 1.0}{0.2}$

whence, $[H^+] = 8 \times 10^{-11}$ g-ion/litre.

Thus, it may be noted that $[H^+]$ falls from 8.94×10^{-6} to 8×10^{-11} g-ion/litre when 1 mole/litre of KCN is added.

Problem 2:

What will be the pH value of a solution obtained by making 5 g of acetic and 7.5 g of sodium acetate, CH_3COONa and making the volume equal to 500 ml. Dissociation constant of acetic acid at 25°C is 1.8×10^{-5}.

Solution:

$$pH = pK_a + \log \frac{[Salt]}{[Acid]}$$

In this case: $pK_a = -\log K_a = -\log (1.8 \times 10^{-5}) = 4.7447$

$$[Salt] = \left[\frac{1000}{500} \times \frac{7.5}{82}\right] \text{ mole per litre}$$

$$= 0.1829 \text{ mole per litre}$$

$$[Acid] = \left(\frac{1000}{500} \times \frac{5}{60}\right) \text{ mole per litre}$$

$$= 0.666 \text{ mole per litre}$$

$$pH = 4.7447 + \log \frac{0.1820}{0.1666}$$

$$= 4.7447 + 0.0435 = 54.7852$$

Thus, $pH = 4.785$

Problem 3:

Calculate the dissociation constant of a weak acid HA, its degree of dissociation of 1 M solution being 10%. What is the pH of this solution?

Solution:

Here $\alpha = 1\%$ or 0.01;

$V = 1$ litre

$$\therefore \quad K_a = \frac{\alpha^2}{(1-\alpha)V} = \frac{(0.01)^2}{(1-0.01)1}$$

$$= 1.01 \times 10^{-4}$$

Now $$[H^+] = \frac{\alpha}{V} = \frac{0.01}{1} = 0.01 \text{ or } 10^{-2}$$

$$\therefore \quad pH = -\log 10^{-2} = 2.0.$$

Problem 4:

What will be the pH of a solution obtained by mixing 800 ml of 0.05 N sodium hydroxide and 200 ml of 0.1 N hydrochloric acid assuming complete ionisation of the acid and the base?

Solution:

$$800 \text{ ml of } 0.05 \text{ N NaOH} \equiv \frac{800}{1000} \times 0.05 \text{ g eqt NaOH}$$
$$= 0.04 \text{ g eqt NaOH}$$

$$200 \text{ ml of } 0.1 \text{ N HCl} \equiv \frac{200}{1000} \times 0.1 \text{ g eqt HCl}$$
$$= 0.02 \text{ g eqt HCl}$$

When 0.04 g eqt NaOH is mixed with 0.02 g eqt HCl, the effective NaOH left

$$= 0.04 - 0.02 = 0.02 \text{ g eqt.}$$

Now total volume = 800 + 200 = 1000 ml = 1 litre

∴ Conc. of NaOH in mixt. = 0.02 g-eqt/litre = 0.02 M

∴ $[OH^-] = 0.02$ g-ion/litre

But $[H^+][OH^-] = 10^{-14}$

∴ $[H^+] = \dfrac{10^{-14}}{0.02} = 5 \times 10^{-12}$ g-ion/litre

∴ $pH = -\log 5 \times 10^{-13}$

$= -\log 5 + \log 10^{13} = -0.6990 + 13.0)$

whence, pH = 11.3010.

Problem 10:

The solubility product of $PbSO_4$ is 1.5×10^{-9} at 298 K. Calculate the solubility of $PbSO_4$ in mol l^{-1}.

Solution:

$PbSO_4$ ionises in solution as follows:

$$PbSO_4(s) \rightleftharpoons Pb^{2+} (aq) + SO_4^{2-} (aq)$$

Let x be the amount of $PbSO_4$ in moles per litre at equilibrium. According to the chemical equations, the concentrations of Pb^{2+} and SO_4^{2-} are follows:

$$[Pb^{2+}] = x$$

$$[SO_4^{2-}] = x$$

$$K_s = [Pb^{2+}][SO_4^{2-}]$$

$$= x \times x$$

or $$1.5 \times 10^{-9} = x^2$$

or $$x = (1.5 \times 10^{-9})^{1/2}$$

$$= (15 \times 10^{-10})^{1/2}$$

$$= 3.875 \times 10^{-5}$$

Problem 5:

The solubility product of lead sulphate is 1.3×10^{-8}. Calculate the number of moles of $PbSO_4$ that can be dissolved in 5 litres of 1×10^{-3} M solution sulphate. Show whether $PbSO_4$ will be precipitated if 20 ml of 2×10^{-4} M $Pb(NO_3)_2$ is mixed with 80 ml of 1×10^{-4} M $PbSO_4$ solution.

Solution:

Let S′ be the solubility of $PbSO_4$ in 1×10^{-3} M Na_2SO_4, then

$$[Pb^{2+}] = S'; [SO_4^{2-}] = (10^{-3} + S')$$

∴ $$K_s = [Pb^{2+}][SO_4^{2-}] = 1.3 \times 10^{-8}$$

or $$S' \times (10^{-3} + S') = 1.3 \times 10^{-8}$$

or $$10^{-3} S' = 1.3 \times 10^{-8}$$ (Neglecting S′ compared to 10^{-2})

or $$S' = \frac{1.3 \times 10^{-8}}{10^{-2}}$$

$$= 1.3 \times 10^{-5} \text{ g mol/litre}$$

∴ No. of moles of $PbSO_4$ dissolved in 5 litre of 1×10^{-3} M Na_2SO_4

$$= 5 \times 1.3 \times 10^{-5} \text{ g mole}$$

$$= 6.5 \times 10^{-5} \text{ g mole}$$

When 20 ml of 2×10^{-4} M $Pb(NO_3)_2$ is mixed with 80 ml of 1×10^{-4} M $PbSO_4$ solution, then

$$[Pb^{2+}] = 0.2 \times 2 \times 10^{-4} + 0.8 \times 1 \times 10^{-4}$$

$$= 1.2 \times 10^{-4} \text{ g-ion/litre;}$$

$$[SO_4^{2-}] = 1 \times 10^{-4} \text{ g-ion/litre}$$

$$\therefore \quad [Pb^{2+}][SO_4^{2-}] = 1.2 \times 10^{-4} \times 1 \times 10^{-4}$$

or $$\text{ionic product} = 1.2 \times 10^{-8}$$

Since the ionic product ($= 1.2 \times 10^{-8}$) is less than its solubility product ($= 1.3 \times 10^{-8}$), so no precipitation of $PbSO_4$ will take place.

Problem 6:

Calculate the H^+ ion concentration of a solution containing 0.5 mole of lactic acid and 0.122 mole of sodium actate dissolved in one litre. (Lactic acid dissociation constant = 1.37 × 10– 4).

Solution:

Here conc. of lactate ion = 0.22 g ion/litre (as the conc. of lactate ion form lactic acid is negligible); Conc. of lactic acid = 0.05 mole/litre.

But $$K_a = \frac{[\text{Lactate ion}][H^+]}{[\text{Lactic acid}]}$$

$$\therefore \quad 1.37 \times 10^{-4} = \frac{[0.122][H^+]}{0.05}$$

whence, $$[H^+] = \frac{1.37 \times 10^{-4} \times 0.05}{0.122}$$

$$= 6 \times 10^{-5} \text{ g-ion/litre}$$

Problem 7:

Using Debye-Huckel limiting law calculate the activity coefficients of sodium ions and of sulphate ions, and the mean-ionic activity coefficient of a 0.01 molal solution of sodium sulphate in water at room temperature.

Solution:

Sodium sulphate is a strong electrolyte which dissociates completely in aqueous solution:

$$Na_2SO_4 \rightarrow 2Na^+ + SO_4^{2-}$$

Thus, the solution is 0.02 molal in Na^+ and 0.01 molal in SO_4^{2-} ions the ionic strength of the solution is given by

$$\mu = \frac{1}{2}(m_1 Z_1^2 \div m_2 Z_2^2)$$

$$= \frac{1}{2}[0.02(1)^2 + 0.01(2)^2] = 0.03$$

According to DHLL equation,

$$\log \gamma Na^+ = -0.509(1)^2(0.03)^{1/2} = -0.088 = \bar{1}.912$$

Taking antilogs,

$$\gamma Na^+ = 0.817$$

Again, $\log \gamma SO_4^{2-} = -0.59(2)^2(0.03)^{1/2} = -0.352 = \bar{1}.648$

Taking antilogs,

$$\gamma SO_4^{2-} = 0.445$$

Hence, the mean-ionic activity coefficient is

$$\gamma\pm = (\gamma^2 Na^+ \gamma SO_4^{2-}) = [(0.817)^2(0.445)]^{1/3} = 0.67.$$

Problem 8:

The molar conductivity at infinite dilution of $AgNO_3$, NaCl and $NaNO_3$ is 116.5, 110.3 and 105.2 ohm^{-1} cm^2 mol^{-1} respectively. The conductivity at AgCl in water is 2.4×10^{-6} ohm^{-1} cm^{-1} and water used in experiment is 1.16×10^{-6} ohm^{-1} cm^{-1}. Find out the solubility of AgCl in g dm^{-3}.

Solution:

$$\lambda_\infty = \frac{1000\ K_{salt}}{C}$$

Here we have

$$(\lambda_\infty)_{AgCl} = (\lambda_\infty)_{AgNO3} + (\lambda_\infty)_{NaCl} - (\lambda_\infty)_{NaNO3}$$

$$= 116.5 + 110.3 - 105.2 = 121.6$$

$$K_{salt} = K_{solution} - KH_2O$$

$$= (2.4 \times 10^{-6} - 1.16 \times 10^{-6})\ ohm^{-1}\ cm^{-1}$$

$$= 1.24 \times 10^{-6}\ ohm^{-1}\ cm^{-1}$$

Substituting these values in the above equation

$$121.6 = \frac{1000 \times 1.24 \times 10^{-6}}{C}$$

or

$$C = \frac{1000 \times 1.24 \times 10^{-6}}{121.6}$$

$$= 1.019 \times 10^{-5}\ \text{equivalent per dm}^3$$

$$= (1.019 \times 10^{-3} \times 143.5) \text{ g per dm}^3$$
$$= 1.4622 \times 10^{-3} \text{ g dm}^{-3}.$$

Problem 9:

The ionic conductance of Na is 50.1 ohm $^{-1}$ cm^2 mol^{-1}. Calculate the ionic mobility of Na^+.

Solution:

Since,

$$\text{Ionic mobility} = \frac{\text{Ionic conductance}}{96500}$$
$$= \frac{(50.1 \text{ ohm}^{-1}\text{cm}^2\text{mol}^{-1})}{(96500 \text{ coulombs mol}^{-1})}$$
$$= 5.1917 \times 10^{-4} \frac{\text{cm}^2}{\text{ohm coulomb}}$$
$$= 5.1917 \times 10^{-4} \frac{\text{cm}^2}{\text{ohm } ampere}$$
$$= 5.1917 \times 10^{-4} \text{ cm}^2 \text{ V}^{-1} \text{ s}^{-1}$$
$$= 5.1917 \times 10^{-84} \text{ m}^{-2} \text{ V}^{-1} \text{ s}^{-1}$$

Problem 10:

At 291 K the ionic velocities of Ag^+ is 0.00057 cm s $^{-1}$ and that of NO_3^- is 0.00063 cm s $^{-1}$. What is the value of λ_∞ for $AgNO_3$ at 291 K? If the specific conductivity of 0.1 N $AgNO_3$ solution is 0.00947 ohm $^{-1}$ cm $^{-1}$, find the degree of dissociation at this dilution.

Solution:

$$\text{Absolute ionic mobility} = \frac{\text{Ionic conductance}}{96500}$$

$$\therefore \quad \lambda_{Ag+} = 0.00057 \times 97500$$
$$\lambda_{NO3--} = 0.00063 \times 96500$$
$$\therefore \quad \lambda_{Ag+} + \lambda_{NO3--} = 96500(0.00057 + 0.00063)$$
$$= 96500 \times 0.0012 = 115.8$$
$$\lambda_c = \frac{100 \times K}{C} = \frac{1000 \times 0.00947}{0.1} = 94.7$$

$$\therefore \qquad \alpha = \frac{\lambda_c}{\lambda_\infty} = \frac{94.7}{115.8} = 0.817.$$

Problem 11:

When a conductance cell was filled with a 0.02 mol dm^{-2} aqueous solution of KCl, which had a specific conductance of 0.2767 S m^{-1}, it had a resistance of 82.39 ohm at room temperature. When the same cell was filled with a 2.5 × 10^{-3} mol dm^{-3} solution of K_2SO_4, its resistance was 325 ohm. Calculate the cell constant and the specific conductance of K_2SO_4 solution.

Solution:

For KCl, $c = 0.02 \text{ mol dm}^{-3} = 0.02 \times 10^3 \text{ mol m}^{-3}$

$R = 82.39$ ohm;

$K = 0.2767 \text{ S m}^{-1}$

$= 0.2767 \text{ ohm}^{-1} \text{ m}^{-1}$

Since K = cell constant/R, we have

Cell constant = KR

$= (0.2767 \text{ ohm}^{-1} \text{ m}^{-1})(82.39 \text{ ohm})$

$= 22.80 \text{ m}^{-1} = 0.2280 \text{ cm}^{-1}$

For K_2SO_4 solution,

K = Cell constant/R

$$= \frac{22.80 \text{ m}^{-1}}{325 \text{ ohm}}$$

$= 7.01 \times 10^{-2} \text{ ohm}^{-1} \text{ m}^{-1}$

$= 7.01 \times 10^{-2} \text{ S m}^{-1}$

Problem 12:

A cell, whose resistance when filled with 0.1 M-KCl solution is 192.3 ohms was found to be 6.306 ohms when filled with 3.186 × 10^{-3} M-NaCl solution at 25°C. Calculate :

(i) the cell constant,

(ii) the specific and equivalent conductivities of NaCl solution, and

(iii) the degree of dissociation of NaCl solution at this dilution. [Given: Sp. conductivity of 0.1 M-KCl is 0.01289 ohm^{-1} cm^{-1}

at 25°C. Equivalent conductivity of solution and chloride ions at 25°C are respective 50.11 and 76.34 ohms^{-1} cm^2.

Solution:

Cell constant, (x) = Sp. Conductivity × Resistance

$$= 0.01289 \text{ ohm}^{-1} \text{ cm}^{-1} \times 192.3 \text{ ohm}$$

$$= 2.479 \text{ cm}^{-1}.$$

Sp. conductivity of 3.186×10^{-3} M-NaCl solution (K)

= Cell Constant/Resistance

$$K = \frac{2.479 \text{ cm}^{-1}}{6.306 \text{ ohm}} = 3.931 \times 10^{-4} \text{ohm}^{-1}\text{cm}^{-1}.$$

But for 3.186×10^{-3} M-NaCl solution,

$$V = \frac{1000}{3.186 \times 10^{-3}} \text{ cm}^2$$

$$\therefore \quad \lambda_r = K \times V$$

$$= \frac{3.931 \times 10^{-4} \times 1000}{3.186 \times 10^{-3}}$$

$$= 123.4 \text{ ohm}^{-1} \text{ cm}^2 \text{ (g eqt)}^{-1},$$

Now $\quad \lambda_\infty \text{ NaCl} = \lambda_{Na+} + \lambda_{Cl-} = 50.11 + 76.34$

$$= 126.45 \text{ ohm}^{-1} \text{ cm}^2 \text{ (g eqt)}^{-1}$$

∴ Degree of dissociation,

$$\alpha = \frac{\lambda_r}{\lambda_\infty} = \frac{123.4}{126.45} = 0.976 \text{ or } 97.6\%.$$

Problem 13:

At 25°C, the specific conductivity of pure water is 5.5×10^{-8} mho cm^{-1}. The ionic conductance of H^+ and OH^- ions at this temperature are 349.8 and 198.5 mhos cm^2 respectively. Calculate the value of the ionisation constant of water.

Solution:

$$\lambda_{\infty\, H2O} = \lambda_{H+} + \lambda_{OH-}$$

$$= 349.8 + 198.5$$

$$= 548.3 \text{ mho cm}^2$$

But conc. of water = 18 g/18 cm^3 = 1 mole/18 cm^3

or $V = 18$ cm^3

$\therefore$ $\lambda_r = K \times V$

$= 5.5 \times 10^{-8} \times 18$

$= 9.9 \times 10^{-7}$ ohm^{-1} cm^2 eqt^{-1}

$$\therefore \quad \alpha = \frac{\lambda_r}{\lambda_\infty}$$

$$= \frac{9.9 \times 10^{-7}}{548.3} \simeq 1.8 \times 10^{-9}$$

It dissociation of water were complete, then concentration of H^+ or OH^- ions would have been 1000/18 g-ion per litre.

But actual degree of dissociation is 1.8×10^{-9}, then

$$[H^+] = [OH^-] = \frac{1000}{18} \times 1.8 \times 10^{-9}$$

$= 10^{-7}$ g-ion/litre

Also $[H_2O] = 1000/18$ g-mole/litre,

$$\therefore \quad K = \frac{[H^+][OH^-]}{[H_2O]}$$

$$= \frac{1 \times 10^{-7} \times 1 \times 10^{-7}}{1000/18}$$

$= 1.8 \times 10^{-16}$.

Problem 14:

A solution of copper (II) sulphate is electrolysed between copper electrodes by a current of 10.0 ampere for 60 minutes. Wheat changes take place at the electrodes and in the solution?

Solution:

The electrode reactions are as follows:

At anode $Cu \rightarrow Cu^{2+} + 2e$

At cathode $Cu^{2+} + 2e \rightarrow Cu$

$$\text{Number of moles of electrons} = \frac{(10.0\ \text{A})(3600\ \text{s})}{(96{,}500\ \text{coulomb / mole})} = 0.373\ \text{F}$$

As it take 2 moles of electrons to react with each mole of copper

(I) or copper metal the number of moles of copper dissolved or deposited is 0.865, *i.e.*,

$$= (0.373 \text{ mole e}) \frac{(1 \text{ mole Cu})}{(2 \text{ moles e})} = 0.1865 \text{ mole Cu}$$

Thus, 0.1865 mole of copper is dissolved from anode, 0.1865 mole of copper is deposited at the cathode, and the original copper (II) ion concentration of the solution remains unchanged.

Problem 15:

When a constant current was passed through a solution of $AuCl_4^-$ ions between gold electrodes for a period of 1.0 minute, the cathode increased in weight by 0.1314 g. How much charge was passed? What was the current, I?

Solution:

At the cathode reduction of gold (III) to gold metal takes place.

$$AuCl_4^- + 3e \longrightarrow Au + 4Cl^-$$

No. of moles of Au

$$= \frac{0.1314 \text{ g}}{197 \text{ g mol}^{-1}} = 0.667 \times 10^{-3} \text{ mol}$$

$$Q = (0.667 \times 10^{-3} \text{ mol Au}) \frac{(3 \text{ mole e})}{(\text{Au mol})}$$

$$= 2.001 \times 10^{-3} \text{ F}$$

$$I = \frac{Q}{t} = \frac{(2.001 \times 10^{-3} \text{ faraday})(96500 \text{ coulomb / faraday})}{60 \text{ seconds}}$$

$$= 3.22 \text{ A.}$$

Problem 16:

Calculate the concentration of OH– ions in 0.1 M sodium acetate solution at 25°C. The ionization constant of acetic acid is 1.8×10^{-5} and $kw = 10^{-14}$ at 25°C.

Solution:

We know that solution,

$$pH = -\log \sqrt{\frac{k_w, K_a}{c}}$$

or $$pH = -\frac{1}{2}\log\frac{k_w K_a}{c}$$

Here $k_w = 10^{-14}$; $K_a = 1.8 \times 10^{-5}$; c = 0.1 M

$$\therefore \quad pH = -\frac{1}{2}\log\frac{10^{-14}\times 1.8\times 10^{-5}}{0.1}$$

$$= \frac{1}{2}\log 18\times 10^{-19} = -\frac{1}{2}\log 1.8\times 10^{-18} = -8.8724$$

$\therefore$ $[H^+]$ = Antilog (– 8.,8724)

= Antilog 9.1276 = 1.342×10^{-8} g-ion/litre

But $k_w = [H^+][OH^-] = 10^{-14}$

$$\therefore \quad [OH^-] = \frac{10^{-14}}{1.342\times 10^{-9}} = 7.454 \times 10^{-5} \text{ g-ion/litre}$$

Problem 17:

Calculate the pH value of 0.001 molar solution of substance B. $K_b = 1.05 \times 10^{-4}$ and $k_w = 1 \times 10^{-14}$.

Solution:

$$pH = -\log\sqrt{\frac{k_w c}{K_b}}$$

$$= \log\sqrt{\frac{K_b}{k_w \times c}} = \frac{1}{2}\{\log K_b - \log k_w - \log c\}$$

Here $K_b = 1.05 \times 14^{-4}$; $k_b = 1 \times 10^{-14}$; c = 0.001 M

$$\therefore \quad pH = \frac{1}{2}\{\log 1.05\times 10^{-5} - \log 1\times 10^{-14} - \log 0.001\}$$

$$= \frac{1}{2}\{-5.0 + 0.0212 + 14.0 + 3.0\} = 6.0105.$$

Problem 18:

Calculate the degree of hydrolysis of 0.5 M solution of sodium acetate at 25°C, given that the dissociation constant of acetic acid is 1.85 × 10^{-5} and ionic product of water is 1.0 × 1.0^{-14} both at the same temperature.

Solution:

Sodium acetate is a salt of *weak acid* and strong base.

$$\therefore \qquad K_h = \frac{k_w}{K_a} = \frac{h^2}{(1-h)V} \simeq \frac{h^2}{V} \quad \text{(if h is small)}$$

or
$$h = \sqrt{\frac{k_w \times V}{K_a}}$$

Now for 0.05 M solution, V = 1/0.05 = 20 litres

$$\therefore \qquad h = \sqrt{\frac{1 \times 10^{-14} \times 20}{1.85 \times 10^{-5}}} = 1.04 \times 10^{-4}.$$

Problem 19:

At 298 K the equivalent conductivity of aniline hydrochloride of 0.01 M solution is 122.5 but in presence of sufficient excess of aniline in practically prevent hydrolysis, it is 106.5 the equivalent conductivity of HCl at infinite dilution is 426.0. Calculate the degree of hydrolysis on the hydrolysis constant.

Solution:

$$C_6H_5NH_3Cl + HOH \rightarrow C_6H_5NH_3OH + HCl$$

If x is the degree of hydrolysis, λ_c is the equivalent conductivity of the hydrolysed salt, λ'_c is the equivalent conductivity of unhydrolysed salt and λ_∞ is the equivalent conductivity of the acid formed, we have

$$x = \frac{\lambda_c - \lambda'_c}{\lambda_\infty - \lambda'_c}$$

$$= \frac{122.5 - 106.5}{426 - 106.5} = \frac{16.0}{319.5} = 0.05$$

and
$$K_h = \frac{x^2 c}{(1-x)} = \frac{0.05 \times 0.05 \times 0.01}{(1-0.05)} = 2.63 \times 10^{-5}.$$